KU-677-529

SI UNITS IN CHEMISTRY

an introduction

R. B. HESLOP

Senior Lecturer in Chemistry
The University of Manchester Institute
of Science and Technology

and

GILLIAN M. WILD

Formerly Head of Chemistry Department
Bury Convent Grammar School

APPLIED SCIENCE PUBLISHERS LTD

LONDON

APPLIED SCIENCE PUBLISHERS LTD
RIPPLE RD., BARKING, ESSEX, ENGLAND

ISBN 85334 515 5

LIBRARY OF CONGRESS CATALOG CARD NUMBER 78-160899

WITH 37 ILLUSTRATIONS AND 26 TABLES

© 1971 APPLIED SCIENCE PUBLISHERS LTD

All rights reserved. No part of this publication may be reproduced, stored in a retrieval system, or transmitted in any form or by any means, electronic, mechanical, photocopying, recording or otherwise, without the prior written permission of the publishers, Applied Science Publishers Ltd, Ripple Road, Barking, Essex, England.

Set in cold type by Plenum Publishing Co. Ltd., Harlesden, London, and printed by Latimer Trend & Co. Ltd., Whitstable, Kent.

Contents

Chapter   Page

Preface. vii

1.   Physical quantities and units. 1
2.   Atomic structure I : the nucleus 13
3.   Atomic structure II : the electron cloud 26
4.   The gas laws : the kinetic theory 39
5.   Ionic crystals. 55
6.   Thermochemistry 70
7.   Chemical equilibria. 81
8.   Chemical kinetics 101
9.   Properties of solutions 116
10.   Ionic equilibria in aqueous solution 130
11.   Oxidation and reduction 146
12.   Electrolytic conduction 158
13.   Dipole moments and electronegativity. 171
14.   Titrations 178
15.   Quantitative analysis of organic compounds . . 193

Appendix. 204

Table of Logarithms 208

Answers 210

Index 223

Preface

This book is written for chemistry students as an introduction to the international system of units (SI) which was first introduced in 1960 and which is becoming widely established. The system certainly simplifies calculation and saves time. Since we feel that full benefit can be gained from the SI only if both the measure and the unit of a physical quantity are substituted into a physical equation, we have introduced the system of calculation known as quantity calculus in Chapter 1 and have used it throughout the book.

The contents of the book have been influenced by recent revisions in the syllabuses of various examining boards, and cover the chemistry required by the 'A' level and scholarship student. The needs of those studying for ONC, HNC and open entrance scholarships have not been neglected, and we hope that this book will be of help to them.

In considering these students we have introduced such subjects as entropy change, dipole moment, activation energy and effective nuclear charge, which should also stimulate the interest of the sixth-form student, even though the quantitative treatment of these topics is outside the syllabus requirements of 'A' level.

We have naturally stressed the quantitative aspects of chemistry and we suggest that the book should be used in conjunction with a general text-book. The examples to the chapters have been specially devised to be arithmetically simple, so that the student can test his understanding rather than his ability to manipulate numbers.

We are extremely grateful to Miss R. McKernon for so capably typing the manuscript.

R. B. Heslop

Gillian M. Wild

Manchester, January 1971.

Chapter 1

Physical quantities and units

Basic physical quantities

In chemistry we are often concerned with the measurement of physical quantities, of which seven (Table I) are generally recognised as independent, basic ones. The choice is not the only one which could have been made; the reasons for it are discussed later in this chapter. Notice that the symbol for a physical quantity is printed in italic (sloping) type.

Derived physical quantities

Other physical quantities are derived from the basic ones by definitions involving only multiplication, division, differentiation or integration, or some combination of these (Table II).

In the fourth column of the table is the <u>dimension</u> of the derived quantity. Since velocity is derived by the differentiation of length with respect to time it is said to have dimension length $\times$ (time)$^{-1}$ and so on.

Throughout this book, derived physical quantities will be defined by the use of equations. The student will see, for example, that the electric resistivity of a conductor is defined by an equation (Eqn. (88) on p. 160) which gives the quantity its correct dimensionality. It is incorrect to define resistivity as the resistance of a cube of material because the dimension of resistivity is different from that of resistance.

The measurement of physical quantities

When we measure the length of an object such as a metal rod we find the ratio of its length to that of some standard

TABLE I

Basic Physical Quantities and the Symbols for Them

Length	l
Mass	m
Time	t
Electric current	I
Thermodynamic temperature	T
Amount of substance	n
Luminous intensity	I_v

length such as a metre scale. Thus, if:

$$\frac{\text{length of rod}}{\text{length of metre scale}} = 0.57$$

the length of the rod is expressed by the equation:

$$l = 0.57 \text{ m}$$

The unit, in this case the metre, is printed in roman (upright) type. The physical quantity is expressed as the product of a number (the <u>measure</u>) and a unit. The equation above can equally well be written:

$$l/\text{m} = 0.57.$$

The printing of numbers

In this book numbers are printed in accordance with international recommendations. The decimal sign is a point on the line. (In some countries a comma is used and this is also accepted internationally). To facilitate the reading of long numbers the digits are grouped in threes on either side of the decimal point but commas are not used, e.g. 8 124.467 25. When a decimal point is placed before the first digit, a zero is placed before the point, e.g. 0.242 65, 0.081.

Very large or very small numbers are usually expressed in forms exemplified by 1.324×10^9 and 3.63×10^{-12}, with a single digit before the decimal point. In cases where, for example, the measure of a quantity between 10^3 and 10^4 is known to only three significant figures, the form 1.23×10^3 is preferred to 1 230, because the latter could be interpreted to mean that the final zero was significant.

TABLE II

Examples of Derived Physical Quantities

Physical Quantity	Symbol	Derivation	Dimension
Velocity	v	$v = \mathrm{d}l/\mathrm{d}t$	length $\times$ (time)$^{-1}$
Acceleration	a	$a = \mathrm{d}v/\mathrm{d}t = \dfrac{\mathrm{d}^2l}{\mathrm{d}t^2}$	length $\times$ (time)$^{-2}$
Force	F	$F = m \times a$	mass $\times$ length $\times$ (time)$^{-2}$
Work	W	$W = \int F \mathrm{d}l$	mass $\times$ (length)$^2 \times$ (time)$^{-2}$
Electric charge	Q	$Q = \int I \mathrm{d}t$	current $\times$ time
Electric potential	V	$V = \mathrm{d}W/\mathrm{d}Q$	mass $\times$ (length)$^2 \times$ (time)$^{-3}$ $\times$ (current)$^{-1}$

Coherent systems of units

If a unit is chosen for each of the independent basic quantities, and all the units for the derived physical quantities are obtained from the basic units by multiplication or division without the introduction of numerical factors, the system of units so derived is said to be coherent. Thus one dyne is the ⟵ unvarying force which, applied to a mass of one gram, produces in it an acceleration of one centimetre per second per second. The dyne belongs to the coherent system of units based on the centimetre as the unit of length, the gram as the unit of mass and the second as the unit of time, the so-called CGS system. In a similar way the newton, that force which will produce in a mass of one kilogram an acceleration of one metre per second per second, belongs to the MKS system which is based on the metre, the kilogram and the second. But the newton is not coherent with the CGS system because

$$1 \text{ newton} = 10^3 \text{ g} \times 10^2 \text{ cm} \times \text{s}^{-2} = 10^5 \text{ dynes}$$

Three-quantity systems of units and their disadvantages

The CGS and MKS systems of units use equations based on only three independent quantities; length, mass and time. These systems are unsatisfactory when applied to magnetism and electricity because some of the physical quantities do not have the same dimension when defined by an equation in electromagnetism as when defined by an equation in electrostatics.

For example, in electrostatics the equation for the force between the two electric charges Q_1 and Q_2 separated by a distance r in a vacuum is given by:

$$F = \frac{Q_1 Q_2}{r^2} \quad \text{(Coulomb's law)} \tag{1}$$

Since F has dimension mass $\times$ length $\times$ (time)$^{-2}$ (Table II) and r has dimension length, Q has dimension (mass)$^{\frac{1}{2}}$ $\times$ (length)$^{\frac{3}{2}}$ $\times$ (time)$^{-1}$ in the three–quantity systems of units.

But in electromagnetism the force between two parallel, linear conductors of length l separated by a distance d in a vacuum and carrying current I_1 and I_2 is given by

$$F = \frac{2I_1 I_2 l}{d} \quad \text{(Ampère's law)} \tag{2}$$

Since F has dimension mass $\times$ length $\times$ (time)$^{-2}$ and both l and d have dimension length, I has dimension (mass)$^{\frac{1}{2}}$ $\times$ (length)$^{\frac{1}{2}}$ $\times$ (time)$^{-1}$. As $Q = \int I \, dt$ (Table II) it must have dimension (mass)$^{\frac{1}{2}}$ $\times$ (length)$^{\frac{1}{2}}$ in the three–quantity system. Thus the dimensionality of Q is not the same when it is defined by an equation in electrostatics as when it is defined by an equation in electromagnetism. It is this discrepancy, in particular, which creates difficulty when three–quantity systems of units like CGS and MKS are used.

Current as a fourth fundamental quantity

One possible way of overcoming the difficulty just outlined is to treat current as an additional fundamental quantity and to introduce proportionality constants into Equations (1) and (2). This solution has been applied, together with another improvement – rationalisation – the introduction of the factor π in situations where there is spherical symmetry and its omission where there is not.

Equations (1) and (2) become:

$$F = \frac{Q_1 Q_2}{4\pi\epsilon_0 r^2} \tag{3}$$

and

$$F = \frac{2\mu_0 I_1 I_2 l}{4\pi d} \tag{4}$$

respectively, where ϵ_0 is the permittivity, and μ_0 the permeability, of a vacuum.

TABLE III
Basic SI Units

Basic Physical Quantity	Name of Unit	Symbol for the Unit
Length	metre	m
Mass	kilogram	kg
Time	second	s
Electric current	ampere	A
Thermodynamic temperature	kelvin	K
Amount of substance	mole	mol
Luminous intensity	candela	cd

If the ampere defined by these equations is to have the same value as the practical unit called the ampere, it becomes necessary to make $\mu_0 = 4\pi \times 10^{-7}$ kg m s^{-2} A^{-2}. For reasons connected with the theory of relativity, which cannot be discussed here, the product $\mu_0 \times \epsilon_0 = c_0^{-2}$, where c_0 is the velocity of light in a vacuum. Thus ϵ_0 has the value $c_0^{-2}/\mu_0 = 8.854 \times 10^{-12}$ kg^{-1} m^{-3} s^4 A^2.

Equations (3) and (4) form the basis of the MKSA (metre, kilogram, second, ampere) system which has been used by electrical engineers for twenty years.

The advantages of further fundamental quantities

It is convenient in thermodynamics to treat temperature as a fundamental physical quantity (i.e., not defined in terms of mass, length, time and current). Similarly it is advantageous in photometry to treat luminous intensity as a fundamental. Recently, physical chemists have suggested the use of amount of substance as a seventh fundamental quantity. Thus the seven quantities shown in Table I have come to be regarded as basic, and the MKSA system of units has been expanded to include seven basic units. Not only are all physical quantities definable in terms of the seven in Table I, but all units can be derived from a basic set of seven, which are given in Table III.

The international system of units

The name Système International (SI) was adopted in 1960 for the system based on these seven units.

The last of the seven (candela) will not concern us further in this book. The other six are now defined as follows:

TABLE IV
Some Derived SI Units with Special Names

Derived Physical Quantity	Name of Unit	Symbol	Definition of Unit
Energy	joule	J	$kg\ m^2\ s^{-2}$
Force	newton	N	$J\ m^{-1}$
Electric charge	coulomb	C	$A\ s$
Electric potential difference	volt	V	$J\ A^{-1}\ s^{-1}$
Electric resistance	ohm	Ω	$V\ A^{-1}$

1. The **metre** is the length equal to 1 650 763.73 wavelengths in vacuum of the radiation arising from the transition between the $2p_{10}$ level and the $5d_5$ level in the ^{86}Kr atom.

2. The **kilogram** is equal to the mass of an international prototype of the kilogram, a piece of platinum – iridium kept at Sèvres. (It may seem surprising that the kilogram has been chosen as the unit of mass rather than the gram. The advantage is that thereby many of the 'practical' electrical units, such as the volt, which have been in use for decades, are made coherent with the system).

3. The **second** is the duration of 9 192 631 770 periods of the radiation corresponding to the transition between the two hyperfine levels of the ground state of the ^{133}Cs atom.

4. The **ampere** is that constant current which, if maintained in two straight, parallel conductors of infinite length, of negligible circular cross-section, and placed in a vacuum one metre apart, would produce between these conductors a force equal to 2×10^{-7} newton per metre of length. (The newton is defined on p. 3).

5. The **kelvin** is 1/273.16 of the thermodynamic temperature (see p. 40) of the triple point of water. (The word degree and the sign ° are obsolescent).

6. The **mole** is that amount of substance which contains the same number of molecules, ions, atoms, electrons, as the case may be, as there are atoms of carbon in exactly 0.012 kg of carbon-12.

The word 'mole' is not synonymous with the older expression 'gram-molecule' which could be applied only to substances

TABLE V
Some Derived SI Units Without Special Names

Physical Quantity	Unit	Symbol
Area	square metre	m^2
Volume	cubic metre	m^3
Density	kilogram per cubic metre	$kg\ m^{-3}$
Pressure	newton per square metre	$N\ m^{-2}$
Molar mass	kilogram per mole	$kg\ mol^{-1}$

which existed as discrete molecules at a specified temperature. A gram-molecule of hydrogen referred to 2.018 g of hydrogen in the form of H_2 molecules. The expression 'a mole of hydrogen' is meaningless, because the word 'mole' can be applied to either an amount of hydrogen atoms or an amount of H_2 molecules. A mole of H_2 obviously has twice the mass of a mole of H. Similarly, a mole of Al_2Cl_6 has twice the mass of a mole of $AlCl_3$. It is always safest to use a formula to specify exactly what the particles are that are being counted.

Derived SI units

The units for several derived physical quantities are given special names in the SI. Those used in this book are shown in Table IV. It will be noticed that units named after scientists are written with small initial letters but the corresponding symbols are capital letters of either the Latin or Greek alphabets.

Table V shows a few examples of derived SI units which are not given special names.

Prefixes for fractions and multiples of SI units

The internationally-agreed prefixes for fractions and multiples of SI units are given in Table VI.

Thus 1 nm is 1×10^{-9} m, 1 kJ is 1×10^3 J. The prefix and the basic symbol are printed close together. It is rather unfortunate that the basic unit of mass itself has a composite symbol (kg). The reason for the choice of this unit has already been explained.

It will be noticed that, apart from the range 10^{-2} to 10^2, all the indices given in the Table are multiples of three. Thus

TABLE VI
Prefixes Used in SI

Fraction	Prefix	Symbol	Multiple	Prefix	Symbol
10^{-1}	deci	d	10	deka	da
10^{-2}	centi	c	10^2	hecto	h
10^{-3}	milli	m	10^3	kilo	k
10^{-6}	micro	μ	10^6	mega	M
10^{-9}	nano	n	10^9	giga	G
10^{-12}	pico	p	10^{12}	tera	T

the measurement of a very large, or a very small, quantity can be expressed as a number between 1 and 1000 if a prefix is used. A typical ionic radius of 1.56×10^{-10} m can be equally well expressed as 156 pm. The former is usually the more convenient for use in calculations but the latter is useful in descriptive writing and for tabulating results.

Summary of the principal changes introduced with the SI

(a) The metre and kilogram replace the centimetre and gram as the coherent units, though the last two remain as sub-multiples (Table VI).

(b) The unit of force is the newton (kg m s^{-2}).

(c) The unit of energy is the joule (kg m^2 s^{-2}); thus the variously defined calories and non-metric units of energy are superseded.

(d) 'Electrostatic' and 'electromagnetic' units are replaced by SI electrical units.

Quantity calculus

The method of calculation used in this book is that of quantity calculus, in which physical quantities, not numbers, are substituted into equations. This is correct, because an equation expresses a physical law which is a relation between physical quantities.

The method can be illustrated by a worked example. Calculate the volume occupied by 2.00 moles of an ideal gas (p. 40) when the pressure is 1.00×10^5 newtons per square metre and the temperature is 300 K.

TABLE VII

The Variation of the Volume of a Gas with Pressure at Constant Temperature

$10^{-5}p/\mathrm{N\,m^{-2}}$	$10^{5}V/\mathrm{m^{3}}$
1.013	2.506
1.331	1.910
1.485	1.712
1.636	1.552

The ideal gas equation is:

$$pV = nRT \qquad (5)$$

Rearranging to find V:

$$V = \frac{nRT}{p}$$

The value of R, the gas constant, to three significant figures, is 8.31 J K^{-1} mol^{-1}.

Substituting physical quantities in the equation:

$$V = \frac{2.00\ \mathrm{mol} \times 8.31\ \mathrm{J\ K^{-1}\ mol^{-1}} \times 300\ \mathrm{K}}{1.00 \times 10^5\ \mathrm{N\ m^{-2}}}$$

$$= \frac{4.99 \times 10^{-2}\ \mathrm{J}}{\mathrm{N\ m^{-2}}}$$

$$= 4.99 \times 10^{-2}\ \mathrm{m^3} \quad (\mathrm{J/N} = \mathrm{m}\ (\text{Table IV}))$$

Note that units are treated as factors in the same way as the numbers. Thus, in the numerator of the fraction above, mol $\times$ mol^{-1} = 1 and K $\times$ K^{-1} = 1. It will usually be necessary, as in the case above, to simplify the units of the answer, but in a coherent system such as SI there is no need to introduce further numerical factors in so doing. A further example will reinforce this point. Calculate the pressure exerted by 3.00 moles of an ideal gas occupying 0.020 0 m^3 at 300 K.

$$p = \frac{nRT}{V}$$

$$= \frac{3.00\ \mathrm{mol} \times 8.31\ \mathrm{J\ K^{-1}\ mol^{-1}} \times 300\ \mathrm{K}}{0.020\ 0\ \mathrm{m^3}}$$

$$= 3.74 \times 10^5\ \mathrm{J\ m^{-3}}.$$

The unit J m^{-3} is not immediately recognisable as a unit of pressure, but since J m^{-1} = N (Table IV), J m^{-3} = N m^{-2}.

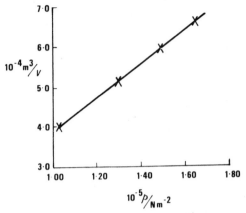

Fig. 1. Variation of V^{-1}
with p at constant tem-
perature.

Thus $p = 3.74 \times 10^5$ N m^{-2}.

Equations involving logarithms

The equation for radioactive decay (p. 21) can be written $n = n_0 e^{-\lambda t}$. n is the amount of a radionuclide left, out of an original amount n_0, after a time t, the decay constant for that nuclide being λ. The equation can equally well be written:

$$\ln \frac{n}{n_0} = -\lambda t. \tag{6}$$

The units of n and n_0 (e.g. mol) cancel, and the logarithm on the left of the equation is, correctly, the logarithm of a number. The form

$$\ln n - \ln n_0 = -\lambda t$$

is not to be recommended because the logarithm of a physical quantity has no meaning. It will be noticed that the dimension of λ must be (time)$^{-1}$ because the left-hand side of Equation (6) is a pure number and the right-hand side must also be dimensionless.

Headings for tables of numerical values

Since the figures placed in the body of a table are numbers, not physical quantities, the headings must also be numbers, and must therefore be written in the form of physical

quantity (in italics) divided by the unit in which the quantity is
measured (e.g. Table VII).

The table indicates that 10^5 V/m^3 = 2.506 when 10^{-5}
p/N m^{-2} = 1.103, i.e. that V = 2.506 × 10^{-5} m^3 when p = 1.103
× 10^5 N m^{-2}, for the amount of gas specified.

The same principle is applied in labelling the axes of a
graph. Since the points on the graph represent numbers (Table
VII), the axes must also be marked off in numbers (Fig. 1).
The graph shows, for example, that 10^{-4} m$^3/V$ = 5.24 when
10^{-5} p/N m^{-2} = 1.331, i.e., that $V = \dfrac{10^{-4} \text{ m}^3}{5.24}$ = 1.910 × 10^{-5} m^3
when p = 1.331 × 10^5 N m^{-2}.

EXAMPLES I

1. Use Table II to work out the dimension of:

   (a) electric resistance (electric potential ÷ current)

   (b) electric capacitance (electric charge ÷ electric
   potential)

   (c) action (work × time).

2. We have seen on p. 9 that the unit J m^{-3} is the same as
N m^{-2}. Use Table IV to express:

   (a) the ohm in terms of kg, m, s and A

   (b) the farad (coulomb ÷ volt) in terms of kg, m, s and A

   (c) the weber (unit of magnetic flux defined as kg m^2 s^{-2}
   A^{-1}) in terms of J and A

   (d) the weber in terms of V and s

3. Use Table IV to complete the following equations by adding
one unit:

   (a) J ÷ C =

   (b) C × Ω = V ×

   (c) J ÷ N =

4. Use Table VI to express:

   (a) 1.326 5 × 10^{-8} m in nm

   (b) 235 g as kg

(c) 1.08×10^{-2} kV as mV

(d) 1 754.8 J as kJ

(e) 1.635×10^{-10} F as pF

taking care to use the conventions on spacing which are explained on p. 2.

5. Calculate the force per cm of length between two parallel straight conductors of infinite length and negligible cross-section placed 1.5 mm apart in a vacuum if a current of 6.0 A flows through one and a current of 4.0 A through the other.

6. Calculate the volume occupied by 2.50 moles of an ideal gas at 500 K and 1.04×10^4 N m^{-2} pressure.

7. Calculate the decay constant for thallium-208 given that $n = 0.250 \, n_0$ when $t = 372$ s (Eqn. 6).

Chapter 2

Atomic structure I: the nucleus

Matter is composed of atoms varying in diameter from about
100-500 pm. To visualise what this size means, imagine a
golf-ball enlarged until it was as big as the Earth. A carbon
atom in it would then become about the size of a golf-ball. The
heaviest naturally-occurring atom (uranium-238) is about 235
times heavier than the lightest (hydrogen-1).

The atomic nucleus

The accepted picture of the atom is of a positively
charged nucleus surrounded by a cloud of negatively charged
electrons. The diameter of the electron cloud is about 10^5
times that of the nucleus; to picture the relative sizes, imagine
a nucleus as big as a matchhead – its electron cloud would fill
a football stadium. The nucleus, however, carries nearly all
the mass, as much as 99.98% in a heavy atom. It consists of
positively charged protons and electrically neutral neutrons
(Table VIII). The atomic number, Z, of an atom is (a) the
number of protons in the nucleus, (b) the number of electrons
in the electron-cloud of the neutral atom, and (c) the number in
the Periodic Table (p. 206) of the element concerned.

Isotopes

The mass number, A, of an atom is the number of
nucleons (protons + neutrons) in the nucleus. A particular
nuclide (*i.e.* a specific kind of atom) is designated by the name
of the element followed by the mass number. Thus lithium-6
is the name given to those atoms of lithium for which $A = 6$
($Z = 3$ and N, the neutron number, $= 3$). Lithium-7, however,
has $Z = 3$ and $N = 4$. Nuclides of the same element, like

13

TABLE VIII

Particles of Some Fundamental Particles

Particle	Symbol	M/m_u	M/kg	$charge/C$
Proton	${}_1^1 p$	1.007 277	$1.672\ 52 \times 10^{-27}$	$+1.602 \times 10^{-19}$
Neutron	${}_0^1 n$	1.008 665	$1.674\ 82 \times 10^{-27}$	0
Electron	e	0.000 548	$9.109\ 1\ \times 10^{-31}$	-1.602×10^{-19}

For the definition of M, the isotopic mass, see below.

lithium-6 and lithium-7, are called isotopes of that element;
isotopes have the same atomic number but different mass
numbers. The symbols are ${}_3^6 Li$ and ${}_3^7 Li$, the mass number and
atomic number being kept to the left to leave space on the right
for charge numbers and atomicities. Thus ${}_8^{16} O_2^{2-}$ refers to a
peroxide ion containing two atoms of oxygen-16 and carrying a
charge equal to that of two electrons; oxygen-16 has $A = 16$ and
$Z = 8$.

Isotopic mass and atomic weight

The isotopic mass of a nuclide, M, is usually expressed
in terms of m_u, the unified atomic mass constant, which is
defined as one-twelfth of the mass of an atom of carbon-12.
The value of m_u is $(1.660\ 43 \pm 0.000\ 08) \times 10^{-27}$ kg. Although
often treated as a unit, it is strictly a physical constant, as is
indicated by the uncertainty in its value.

Isotopic masses can be measured with great accuracy in an
instrument called a mass spectrograph. For ${}_3^6 Li$, $M = 6.015\ 13\ m_u$
and for ${}_3^7 Li$, $M = 7.016\ 01\ m_u$. The atomic weight, more correctly
called the relative atomic mass, A_r, of an element is a number; it
is the weighted mean of the masses of the naturally-occurring isotopes divided by m_u. For lithium, which contains 7.42% of ${}_3^6 Li$ and
92.58% of ${}_3^7 Li$, $A_r = [(6.015\ 13\ m_u \times 0.074\ 2) + (7.016\ 01\ m_u$
$\times 0.925\ 8)] \div m_u = 6.941\ 8$.

Non-additivity of mass: nuclear binding energy

It might be expected that the mass of an atom would be
the sum of the masses of the protons, neutrons and electrons
contained in it. In fact, all atoms except hydrogen-1 weigh less
than their constituent parts. The reason is that there is a
large decrease in potential energy which accompanies the

binding of the protons and neutrons, collectively called nucleons, in the nucleus.

The Einstein equation:

$$\Delta E = \Delta m \, c_0^2 \tag{7}$$

where ΔE is the energy release, Δm is the loss of mass and c_0 is the velocity of light in a vacuum, expresses the relation between energy and mass. Strictly, the mass of a particle depends on its velocity, but it is sufficient here to equate the masses of the protons and neutrons to their rest masses, i.e., their masses at zero velocity.

The difference between the mass of a $_3^7\text{Li}$ atom and the protons, neutrons and electrons from which it is formed is calculated below.

Total mass of 3 protons and 3 electrons
(i.e.) $3\,^1\text{H}$ atoms, $= 3 \times 1.007\,825\,m_u$ $= 3.023\,48\,m_u$

Total mass of 4 neutrons $= 4 \times 1.008\,665\,m_u$ $= 4.034\,66\,m_u$

Total mass of 3 protons, 4 neutrons and
3 electrons $= 7.058\,14\,m_u$

Isotopic mass of $_3^7\text{Li}$ $= 7.016\,01\,m_u$

Difference $= 0.042\,13\,m_u$

Using Einstein's equation (Eqn. 7), the mass is equivalent to the energy ΔE:

$$\Delta E = 0.042\,13 \times 1.660 \times 10^{-27} \text{ kg} \times (2.998 \times 10^8 \text{ m s}^{-1})^2,$$

since $1\,m_u = 1.660 \times 10^{-27}$ kg

and $c_0 = 2.998 \times 10^8$ m s^{-1} (Table XXIII).

Thus $\Delta E = 6.29 \times 10^{-12}$ kg m^2 s^{-2}

$= 6.29 \times 10^{-12}$ J (Table IV)

This quantity is the binding energy of the $_3^7\text{Li}$ nucleus, ⟵ the energy released when it is formed from its protons and neutrons, or, alternatively, the energy which would be needed to split the nucleus itself into separate nucleons. The binding energy per nucleon for $_3^7\text{Li}$ is therefore $(6.29 \times 10^{-12}$ J$)/7$ nucleons $= 8.98 \times 10^{-13}$ J/nucleon.

The variation of binding energy per nucleon with mass number

The variation of $B.E.$ per nucleon with mass number is shown as a smooth curve in Fig. 3. Nuclei with mass numbers around 60 are those with the greatest $B.E.$/nucleon.

Analysis of this curve leads to the empirical formula

$$10^{13} \times B.E. \text{ per nucleon}/J = 22.6 - 20.8\, A^{-\frac{1}{3}} - 0.96\, Z^2\, A^{-\frac{4}{3}} \quad (8)$$

The first term represents the binding energy of a nucleon with twelve others packed closely round it, the second term represents the loss of $B.E.$ due to the fact that some of the nucleons are at the surface and are incompletely surrounded by others, the final term represents the loss of $B.E.$ due to the coulombic repulsion between protons. For nuclides of small mass the second term is proportionately large, for nuclides of large mass and charge the final term is of increased importance. Hence the maximum values for $B.E.$ per nucleon occur at intermediate values of A.

Fission and fusion reactions

The process known as <u>nuclear fission</u> is the source of energy in nuclear power stations. The most commonly-used process is one in which uranium-235 nuclei break up when hit by neutrons, each producing two large nuclei and additional neutrons, which perpetuate the reaction chain by attacking other U-235 nuclei. A typical nuclear reaction in a nuclear reactor would be:

$$^{235}_{92}U + ^{1}_{0}n \rightarrow \text{compound nucleus of very short life}$$

$$\rightarrow ^{95}_{39}Y + ^{138}_{53}I + 3\,^{1}_{0}n$$

Note that in a nuclear reaction, the sum of the mass numbers remains unchanged ($235 + 1 = 95 + 138 + 3$) and the sum of the atomic numbers also remains unchanged ($92 + 0 = 39 + 53 + 0$). But there is a reduction of mass, and energy is set free in accordance with Einstein's equation.

Primary fission products such as the ^{95}Y and ^{138}I above undergo radioactive decay (p. 18) to give stable nuclides such as ^{95}Mo and ^{138}La. If the isotopic masses of these nuclides are known, it is possible to calculate the total energy which should be available from the fission process.

$$M\left(^{235}_{92}U\right) \qquad\qquad = 235.043\ 9\ m_u$$

$$M\left(^{1}_{0}n\right) \qquad\qquad = \quad 1.008\ 7\ m_u$$

$$\overline{236.052\ 6\ m_u}$$

$$M\left(^{95}_{42}Mo\right) \qquad\qquad = \quad 94.904\ 6\ m_u$$

$$M\left(^{138}_{57}La\right) \qquad\qquad = 137.906\ 8\ m_u$$

$$3\ M\left(^{1}_{0}n\right) \qquad\qquad = \quad \underline{3.026\ 1\ m_u}$$

$$235.837\ 5\ m_u$$

Thus the loss of mass is $0.215\ 1\ m_u$. The mass lost by the fission of one mole of uranium-235 ($0.235\ 043\ 9$ kg) would be $0.215\ 1 \times 10^{-3}$ kg, which is equivalent to

$$0.215\ 1 \times 10^{-3}\ kg \times (2.998 \times 10^8\ m\ s^{-1})^2$$
$$= 1.935 \times 10^{13}\ kg\ m^2\ s^{-2} = 1.935 \times 10^{13}\ J$$

The ordinary domestic unit of electricity, the kilowatt-hour (kWh), is 3.6×10^6 J. Thus 0.235 kg of U-235 can release more than five million kWh through the process of nuclear fission, sufficient energy to supply four hundred families with electricity for a year.

The shape of the curve in Fig. 2 indicates that an increase in B.E., and therefore a release of energy, should occur not only when a very heavy nucleus breaks into nuclei of moderate size but also when very small nuclei fuse together, if they can be made to do so. In fact there is strong evidence that our sun and the other stars produce energy by nuclear fusion reactions. Unfortunately, the conditions under which these reactions occur, temperatures of 10^8-10^9 K, have so far not been achieved experimentally for anything but very short periods. One obvious difficulty is that of keeping material at that temperature without vaporising the walls of the container. However, the advantages which would accrue from the success-ful technological exploitation of nuclear fusion are tremendous, as can be seen from the following calculation.

Consider the possible fusion reaction:

$$2\ ^{2}_{1}H \longrightarrow\ ^{4}_{2}He$$

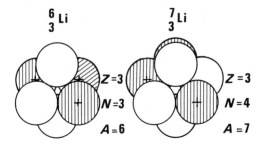

In the neutral atom the electron cloud contains three electrons in each case

Fig. 2.   Representation of the nuclei of
^6_3Li and ^7_3Li.

$$2\,M\,(^2_1\text{H}) \;=\; 4.028\ 20\ m_\text{u}$$

$$M\,(^4_2\text{He}) \;=\; 4.002\ 60\ m_\text{u}$$

Energy
release $= 0.025\ 60\ m_\text{u}$

The fusion of two moles of ^2_1H (i.e. 0.004 028 20 kg) to form helium–4 would therefore release

$$0.025\ 60\ \times 10^{-3}\ \text{kg} \times (2.998 \times 10^8\ \text{m s}^{-1})^2$$
$$= 2.304 \times 10^{12}\ \text{J}$$

This is the equivalent of 6.4×10^5 kWh, sufficient energy to supply a one-bar electric fire for more than seventy years. It is hardly surprising that physicists are interested in achieving successful, controlled, nuclear fusion reactions.

Radioactive decay

Many of the heaviest nuclides, such as uranium–238 and thorium–236, undergo spontaneous changes in which particles are released from the nucleus. The most common of these radioactive decay processes are α–emission:

$$\text{e.g.}\quad ^{238}_{92}\text{U} \longrightarrow ^{4}_{2}\text{He} + ^{234}_{90}\text{Th}$$

(in which the nucleus emits an α–particle, a $^4_2\text{He}^{2+}$ ion consisting of a helium–4 nucleus containing two protons and two neutrons),

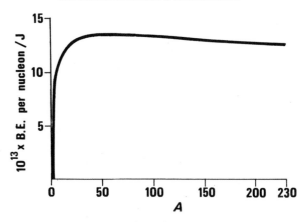

Fig. 3. Variation of binding energy per nucleon with mass number: smoothed curve.

and β^--emission:

e.g. $^{234}_{90}\text{Th} \longrightarrow {}^{234}_{91}\text{Pa} + {}^{0}_{-1}\beta$

(in which a β-particle, with the properties of an electron, is released with high kinetic energy from the nucleus).

The daughter nuclide obtained in an α-emission (e.g. $^{234}_{92}\text{Th}$ above) has A less by four units and Z less by 2 units than the parent nuclide. A β^--emission produces a daughter nuclide with the same mass number as the parent but with one more unit of charge on the nucleus. Thus an α-emission followed by two β^--emissions gives rise to a nuclide isotopic with the original one:

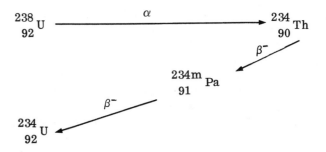

The nuclides $^{234}_{90}\text{Th}$, $^{234}_{91}\text{m Pa}$ and $^{234}_{92}\text{U}$ are isobaric, i.e., their mass numbers are the same. The letter m attached to

TABLE IX

Isotopic Masses and Abundance of Stable Nuclides, and Relative Atomic Masses of Some Light Elements.

Nuclide	M/m_u	Abundance $\times 10^2$	A_r for Element
^{1}H	1.007 825	99.985	1.007 97
^{2}H	2.014 10	0.015	
^{4}He	4.002 60	~100	4.002 60
^{6}Li	6.015 13	7.42	6.939
^{7}Li	7.016 01	92.58	
^{9}Be	9.012 19	100	9.012 19
^{10}B	10.012 94	19.6	10.811
^{11}B	11.009 31	80.4	
^{12}C	12.000 00	98.89	12.011 15
^{13}C	13.003 35	1.11	
^{14}N	14.003 07	99.63	14.006 7
^{15}N	15.000 11	0.37	
^{16}O	15.994 91	99.759	15.999 4
^{17}O	16.999 14	0.037	
^{18}O	17.999 16	0.204	
^{19}F	18.998 40	100	18.998 40

the mass number of $^{234\,m}_{91}$Pa indicates that this particular nuclide is produced in a metastable (excited) state. There is an isomer, $^{234}_{91}$Pa, of much longer half-life, but it is not produced by the decay of thorium-234. The three changes shown above are the first three steps in the <u>uranium series</u>, a naturally-occurring series of nuclear disintegrations by which uranium-238 decays to lead-206, a stable nuclide which does not disintegrate further. An α- or β^--emission is, in some cases, accompanied by the release of energy in the form of a γ-ray, which has the properties of a very penetrating X-ray. A γ-ray is produced when the daughter nuclide formed by the loss of an α- or β^--particle is produced with its nucleus in an excited state; the nucleons within it rapidly rearrange themselves in their lowest energy levels as energy is lost in the form of a γ-ray.

Decay constant: half-life

The rate at which the amount of a radionuclide decreases as a result of radioactive decay is proportional to the amount:

$$\frac{-\mathrm{d}n}{\mathrm{d}t} = \lambda n \tag{9}$$

The proportionality constant, λ, with dimension $(\text{time})^{-1}$, is the <u>decay constant</u> for the process. ⬅

Integrating Eqn. (9):

$$\frac{n}{n_0} = \exp(-\lambda t) \tag{10}$$

where n is the amount of nuclide remaining out of an original amount n_0 when an interval, t, has elapsed. Figure 4 illustrates the variation of the ratio n/n_0 with time.

The time in which the amount of nuclide is reduced to one-half of the original amount is called the <u>half-life</u>, $t_{\frac{1}{2}}$, of ⬅ the nuclide. Its relation to λ is obtained as follows:

$$\exp(-\lambda t_{\frac{1}{2}}) = \frac{n}{n_0} = \frac{1}{2}$$

$$\therefore \; -\lambda t_{\frac{1}{2}} = \ln\frac{1}{2}$$

$$\lambda t_{\frac{1}{2}} = \ln 2 = 0.693$$

$$t_{\frac{1}{2}} = 0.693/\lambda \tag{11}$$

For $^{234\,m}\text{Pa}$ (p. 19), $t_{\frac{1}{2}} = 70$ s

$$\therefore \; \lambda = \frac{0.693}{70 \text{ s}} = 9.9 \times 10^{-3} \text{ s}^{-1}.$$

The value of $t_{\frac{1}{2}}$ for a radionuclide varies greatly from one to another: it can be thousands of millions of years (e.g. $^{238}_{92}\text{U}$, $t_{\frac{1}{2}} = 4.5 \times 10^9$ years) or a small fraction of a second (e.g. $^{214}_{84}\text{Po}$, $t_{\frac{1}{2}} = 1.6 \times 10^{-4}$ s).

The <u>activity</u> (a) of a specimen of a radionuclide is mea- ⬅ sured in curies. The curie (Ci) is not an SI unit, but is defined as the amount in which 3.70×10^{10} radioactive disintegrations are occurring per second. The half-life of a nuclide can be calculated from the rate at which the activity of a specimen decreases. In comparing activities it is sufficient to compare the number of disintegrations recorded in a counting device such as a Geiger-Muller counter. Suppose a specimen of a radionuclide gives a count-rate of 131 counts per second originally and a count-rate of 71 counts per second after 30 minutes.

$$\text{Since } \ln\frac{n_0}{n} = \lambda t \tag{12}$$

$$2.303 \log_{10} \frac{131}{71} = \lambda \times 30 \times 60 \text{ s}$$

$$\lambda = 3.4 \times 10^{-4} \text{ s}^{-1}$$

and $$t_{\frac{1}{2}} = \frac{0.693}{\lambda} = 2.04 \times 10^3 \text{ s}$$

It should be noted that both λ and $t_{\frac{1}{2}}$ for a particular nuclide are independent of mass. If 1.0 mg of a nuclide of half-life 70 s decays for 140 s the mass of that nuclide remaining will be 0.25 mg; if 1.0 kg decays for 140 s the mass of that nuclide remaining will be 0.25 kg.

Radiocarbon dating

An interesting use of Eqn. (10) is in estimating the age of archeological specimens. The carbon in living matter contains a definite proportion of $^{14}_{6}C$ ($t_{\frac{1}{2}}$ = 5580 years) which is formed by the action of neutrons in cosmic rays upon the nitrogen-14 in the atmosphere:

$$^{1}_{0}n + ^{14}_{7}N = ^{1}_{1}H + ^{14}_{6}C$$

The radiocarbon is soon converted into CO_2 which is taken up by plants in photosynthesis and made into carbohydrates to be consumed by animals who return some of it to the atmosphere in the course of respiration. As a result of the familiar plant-animal carbon cycle there is a constant equilibrium concentration of $^{14}_{6}C$ present in living matter. An examination of specimens of new wood from sources in different parts of the world indicates that there is a constant radioactivity in it equivalent to 15.3 counts per minute per gram of carbon.

Once a carbon-containing material is separated from equilibrium with the living carbon cycle the $^{14}_{6}C$ in it decays and the activity of the specimen decreases. Thus, for example, a piece of wood which was cut from a tree a thousand years ago for use in the building of a Viking longship or a

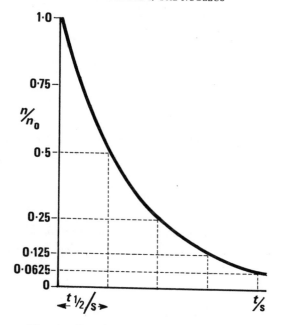

Fig. 4. Variation of n/n_0 with time.

Saxon church has a radioactivity less than that of 'living' carbon. Its activity α, is given by:

$$\frac{\alpha}{15.3 \text{ min}^{-1}} = \exp(-\lambda t) = \exp\left(-\frac{0.693 \times 1000\text{y}}{5580\text{y}}\right)$$

from which $\alpha = 13.5 \text{ min}^{-1}$.

In practice, dating is carried out by comparing the count-rate of carbon prepared from new wood with that of the old carbon under identical circumstances in the same counting equipment. (Question 9, (p. 24).

EXAMPLES II

1. Calculate A_r for magnesium from the following data:

	M/m_u	Abundance $\times 10^2$
^{24}Mg	23.985 04	78.70
^{25}Mg	24.985 84	10.13
^{26}Mg	25.982 59	11.17

2. Calculate A_r for silicon from:

	M/m_u	Abundance $\times 10^2$
^{28}Si	27.976 93	92.21
^{29}Si	28.976 49	4.70
^{30}Si	29.973 76	3.09

3. Use Tables VIII and IX to calculate the *B.E.* of (a) the ^{12}C nucleus, (b) the ^{13}C nucleus. Calculate also the *B.E.* per nucleon in each case.

4. Use Tables VIII and IX to calculate the *B.E.* and *B.E.* per nucleon for (a) ^{10}B and (b) ^{11}B.

5. Use Eqn. (8) to calculate the *B.E.* per nucleon for (a) ^{27}Al and (b) ^{64}Cu.

6. Balance the following nuclear equations by adding one further species.

(a) $\quad ^{27}_{13}Al \;+\; ^{1}_{0}n \;=\; ^{4}_{2}He \;+$

(b) $\quad ^{9}_{4}Be \;+\; ^{4}_{2}He \;=\; ^{12}_{6}C \;+$

(c) $\quad ^{35}_{16}S \;=\; ^{0}_{-1}\beta \;+$

7. Calculate the energy released in the reactions:

(a) $\quad ^{14}_{7}N \;+\; ^{1}_{0}n \;=\; ^{14}_{6}C \;+\; ^{1}_{1}H$

(b) $\quad ^{6}_{3}Li \;+\; ^{1}_{0}n \;=\; ^{4}_{2}He \;+\; ^{3}_{1}H$

For $^{14}_{6}C$, $M = 14.003\ 24\ m_u$, for $^{3}_{1}H$, $M = 3.016\ 05\ m_u$. The other isotopic masses are given in Tables VIII and IX.

8. A sample of a radionuclide, which gives a count-rate of 680 s^{-1} when first placed in a counting device, is found to give a count-rate of only 125 s^{-1} after 350 s. Calculate (a) λ (b) $t_{\frac{1}{2}}$ for the nuclide.

9. The β^--activity from 1.0 g of carbon made from the wood of a recently-cut tree registered a count-rate of 0.204 s^{-1}. A one-gram specimen of carbon prepared from wood taken from a

Viking longship gave a count-rate of 0.177 s^{-1}. Estimate the age of the ship to the nearest 50 years. For ^{14}C, $t_{\frac{1}{2}} = 5580$ years.

Chapter 3

Atomic structure II: the electron cloud

Electromagnetic waves

The idea that light was a wave-motion arose out of the attempt
to explain the diffraction of light rays, and all radiant energy
is conveniently described as being propagated in the form of
waves. Figure 5 illustrates the meaning of the term wavelength
applied to a continuous train of waves.

Radio waves, heat rays, light rays, ultraviolet rays, X-
rays and γ-rays are all examples of electromagnetic waves.
They differ in wavelength, λ, as shown in Fig. 6.

Experiments show that these radiations all travel, in a
vacuum, with the same velocity, c_0, = 2.998 × 10^8 m s^{-1}.

The frequency of the radiation, ν, is related to the wavelength by

$$\nu = c_0/\lambda \tag{13}$$

Frequency has the dimension (time)$^{-1}$ and is expressed in the
SI as s^{-1}, a unit also known as the hertz (Hz).

Another useful concept is that of wave number, which is λ^{-1} for
the radiation concerned. It is equal to ν/c_0 and has dimension
(length)$^{-1}$.

Photons

Although the diffraction of electromagnetic waves is best
explained in terms of a wave picture, there are other phenomena which can only be explained satisfactorily by imagining a ray
to consist of particles. One of these phenomena is the photoelectric effect, which concerns the emission of electrons by a
surface irradiated with ultraviolet or visible rays. The clean

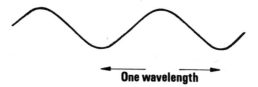

Fig. 5. The meaning of wavelength.

surface of an alkali metal is particularly suitable for these experiments.

Figure 7 shows the relation between the frequency, ν, of monochromatic ultraviolet or visible radiation shining on a metal surface and the kinetic energy, E, of the electrons emitted from the surface. Electrons are not emitted at all unless the incident radiation has a certain minimum threshold frequency, ν_0, characteristic of the metal. Radiation of higher frequency, ν, causes the emission of electrons; the kinetic energy with which they leave the surface, measured by finding the opposing voltage which is just sufficient to bring them to a stop, is proportional to $\nu - \nu_0$.

$$E = \text{a constant} \times (\nu - \nu_0) \qquad (14)$$

The constant, obtained from the slope of the graph in Fig. 7, is Planck's constant, h, $= 6.6256 \times 10^{-34}$ J s. Notice that the electrons are given an energy which depends on the frequency, and not on the intensity, of the incident radiation.

It was suggested by Einstein that the result could be explained by imagining that a single particle of radiant energy, a photon, transmitted all of its energy to a single electron in the metal; the electron then had a certain amount of work to do (a constant for a particular metal is known as its work function) to escape from the surface. The energy remaining to the electron is the kinetic energy with which it escapes. Rearranging Eqn. (14):

$$h\nu = h\nu_0 + \tfrac{1}{2}m_e v^2 \qquad (15)$$

| (energy of incident photon) | (work function of the metal) | (kinetic energy of the electron emitted) |

Notice that the equation implies an energy transfer from one photon to one electron, not a dissipation among many electrons. Furthermore, the energy of a photon is proportional to

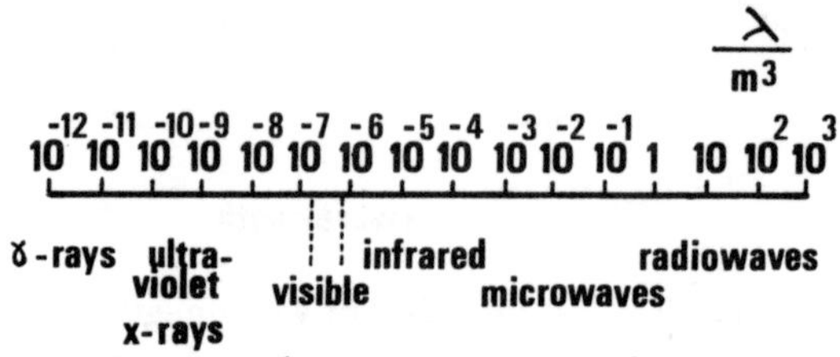

$$\frac{\lambda}{m3}$$

| -12 | -11 | -10 | -9 | -8 | -7 | -6 | -5 | -4 | -3 | -2 | -1 | | 2 | 3 |

$$10^{-12}\ 10^{-11}\ 10^{-10}\ 10^{-9}\ 10^{-8}\ 10^{-7}\ 10^{-6}\ 10^{-5}\ 10^{-4}\ 10^{-3}\ 10^{-2}\ 10^{-1}\ 1\ 10\ 10^{2}\ 10^{3}$$

ɣ-rays ultra-   infrared   radiowaves
          violet visible   microwaves
x-rays

Fig. 6.   Electromagnetic radiation.

the frequency of the radiation of which it forms a part:

$$E_{photon} = h\nu_{photon} \qquad (16)$$

We are forced to the disturbing conclusion that electromagnetic radiation has some of the properties of a wave and some of the properties of a stream of particles – disturbing in the sense that, from our experience of larger bodies, we cannot form a simple physical picture of a particle with wave properties.

The free electron as a wave

The idea that if a wave could behave like a particle, a particle might also behave like a wave, was suggested by de Broglie, who used Eqns. (7), (13) and (16) to derive the equation:

$$\lambda = h/mv \qquad (17)$$

where λ is the wavelength characteristic of a particle of mass m moving with velocity v.

Davisson and Germer verified this equation for the free electron by an experiment in which electrons, which were accelerated through a potential of 54 V, and passed through powdered crystals, produced a diffraction pattern characteristic of X-rays of wavelength 165 pm.

Theoretical calculation of the wavelength of a free electron

If an electron of charge e is accelerated through a potential difference V the energy acquired is Ve. Thus, an electron with charge 1.60×10^{-19} C accelerated through a potential difference of 54 V, as in the experiment described above, emerges with a kinetic energy E:

$$
\begin{aligned}
E = Ve &= 54\ V \times 1.60 \times 10^{-19}\ C \\
&= 8.65 \times 10^{-18}\ C\ V \\
&= 8.65 \times 10^{-18}\ J \qquad \text{(Table IV)}
\end{aligned}
$$

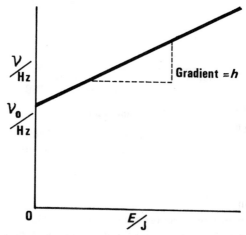

Fig. 7. Photoelectric effect: Variation of the energy of the emitted electrons with frequency of incident radiation.

The mass of the electron, m_e, is 9.11×10^{-31} kg (Table VIII). Since its kinetic energy $= \frac{1}{2} m_e v^2 = 8.65 \times 10^{-18}$ J,

$$v^2 = \frac{2 \times 8.65 \times 10^{-18} \text{ kg m}^2 \text{ s}^{-2}}{9.11 \times 10^{-31} \text{ kg}}$$

and

$$v = 4.35 \times 10^6 \text{ m s}^{-1}.$$

Using Eqn. (17), in which Planck's constant, h, $= 6.63 \times 10^{-34}$ J s

$$\lambda = \frac{h}{m_e v} = \frac{6.63 \times 10^{-34} \text{ kg m}^2 \text{ s}^{-1}}{9.11 \times 10^{-31} \text{ kg} \times 4.35 \times 10^6 \text{ m s}^{-1}}$$

$$= 1.67 \times 10^{-10} \text{ m}$$

$$= 167 \text{ pm}$$

This is in reasonable agreement with the experimental result of 165 pm.

The atomic spectrum of hydrogen

The passage of an electric discharge through hydrogen causes the gas to glow. If the light which is emitted is dispersed through a prism it gives rise to a spectrum, the atomic spectrum of hydrogen, consisting of sharp lines (Fig. 8). ←

In 1885, Balmer pointed out that the wave numbers of the lines were given by the formula

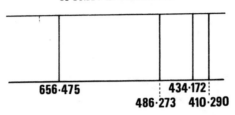

656·475   434·172
486·273   410·290

Wavelength / nm

Fig. 8. First four lines in the visible
range in the atomic spectrum of hydrogen.

$$\frac{1}{\lambda} = \text{a constant } (\frac{1}{2^2} - \frac{1}{n^2})$$ (18)

where n is an integer greater than 2. For $n = 3$, $\lambda = 656.475$ nm,
for $n = 4$, $\lambda = 486.273$ nm, and so on. The constant is the Rydberg constant for hydrogen, R_H, equal to $1.096\ 775 \times 10^7\ \text{m}^{-1}$.

Subsequent experiments showed that there were other
series of lines with similar equations connecting them, all of
the form:

$$\frac{1}{\lambda} = R_H\ (\frac{1}{n_1^2} - \frac{1}{n_2^2})$$ (19)

where $n_2 > n_1$. The series in which $n_1 = 1$ is the Lyman series
of lines in the ultraviolet; the series in which $n_1 = 3$ is the
Paschen series in the infrared.

Thus, for the first line in the Lyman series:

$$\frac{1}{\lambda} = 1.096\ 775 \times 10^7\ \text{m}^{-1}\ (\frac{1}{1^2} - \frac{1}{2^2})$$

$$= 0.822\ 68 \times 10^7\ \text{m}^{-1}.$$

$$\therefore \lambda = 1.215\ 5 \times 10^{-7}\ \text{m} = 121.55\ \text{nm}$$

and, for the first line in the Paschen series:

$$\frac{1}{\lambda} = 1.096\ 775 \times 10^7\ \text{m}^{-1}\ (\frac{1}{3^2} - \frac{1}{4^2})$$

$$= 5.331 \times 10^5\ \text{m}^{-1}$$

$$\therefore \lambda = 1.876 \times 10^{-6}\ \text{m} = 1.876\ \mu\text{m}.$$

An explanation of the hydrogen spectrum was offered in
1913 by Bohr on the basis of a model in which a proton and an

electron rotated about their centre of gravity. However, a
similar rotating model containing a nucleus and more than one
electron was unsatisfactory for explaining the atomic spectra
of elements with $Z > 1$. A more satisfactory model, based on
the concept of the bound electron as a wave, was developed in
1925 by Schrödinger.

The bound electron as a wave

Schrödinger used the idea that if the free electron could
be considered to have the properties of a travelling wave, an
electron in an atom such as that of hydrogen might be considered to have the properties of a standing wave, rather like
the wave of a vibrating guitar string but, in this case, three-
dimensional in character. Possible modes of vibration of a
guitar string are limited because the string is fixed at its ends,
similarly the possible modes of vibration of the electron wave
are limited by the fact that the electron is bound to the nucleus.

These restrictions quite naturally lead to the result that,
in the case of the hydrogen atom, for example, there are only
certain possible values for the energy of the whole system of
proton and of proton and bound electron. These are

$$-2.180 \times 10^{-18} \text{ J}, \quad \frac{-2.180}{2^2} \times 10^{-18} \text{ J}, \ldots \ldots \frac{-2.180}{n^2} \times 10^{-18} \text{ J}.$$

The energy of the system is said to be quantised. It should
be noted that the energies are negative, indicating that the
electron is bound to the proton.

Relation between the atomic spectrum and the energy levels in the hydrogen atom

Figure 9 illustrates the possible energy levels in a
hydrogen atom and the transitions which produce the lines in
the atomic spectrum. If an atom in its lowest energy (ground)
state is energised in an electric discharge and raised to, say,
the second excited state, and it subsequently loses energy to
fall to the first excited state, the energy release is:

$$E_3 - E_2$$

$$= \left\{ \frac{-2.180}{9} - (-\frac{2.180}{4}) \right\} \times 10^{-18} \text{ J}$$

$$= 2.180 \times 10^{-18} \text{ J } (\frac{1}{2^2} - \frac{1}{3^2}).$$

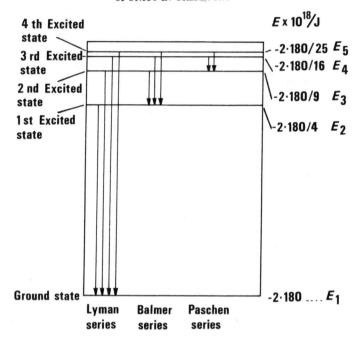

Fig. 9. Energy levels in the hydrogen atom.

If the energy of the transition is released as a photon with the energy $E_3 - E_2 = h\nu = hc_0/\lambda$, the wave number of the line produced in the spectrum should be:

$$\frac{1}{\lambda} = \frac{2.180 \times 10^{-18} \text{ J}}{hc_0} \left(\frac{1}{2^2} - \frac{1}{3^2}\right)$$

$$= \frac{2.180 \times 10^{-18} \text{ J}}{6.625\,6 \times 10^{-34} \text{ J s} \times 2.998 \times 10^8 \text{ m s}^{-1}} \left(\frac{1}{2^2} - \frac{1}{3^2}\right)$$

$$= 1.096\,8 \times 10^7 \text{ m}^{-1} \left(\frac{1}{2^2} - \frac{1}{3^2}\right)$$

Note that the constant is the Rydberg constant for hydrogen (p. 30). Thus the treatment of the bound electron as a wave has led quite naturally to the formula for the spacing of the lines in the hydrogen spectrum, incorporating the integers first noticed by Balmer.

Figure 9 illustrates how the quantised energy levels of the hydrogen atom 'crowd together' in the higher excited states. It is also clear that the lines in the Lyman series are produced by much larger energy transitions than those in the Balmer series which, in turn, are produced by larger transitions than those in the Paschen series.

Ionisation energy

It can be seen from Fig. 9 that if a hydrogen atom in its ground state is provided with 2.180×10^{-18} J the energy of the system becomes zero, indicating that the electron is no longer bound to the nucleus. The energy needed to achieve this is the ionisation energy, I, of the hydrogen atom. For other atoms, the first ionisation energy refers to the energy needed to remove the least strongly bound electron from the ground-state atom.

Atomic orbitals in hydrogen

Schrödinger's treatment of the bound electron as a three-dimensional wave has led to a representation of an electron with a particular energy as having a certain probability of being in a small element of volume at a certain position relative to the nucleus. The density of dots around the nucleus in Fig. 10 represents the probability of finding an electron at a particular position in the ground-state hydrogen atom. It is sufficient for most purposes to represent as a sphere the volume within which the electron spends 95% of its time. This is called the 1s orbital of the hydrogen atom.

For a hydrogen atom in the first excited state there are four different orbitals which describe the possible electron probability distribution. One of these, the 2s orbital, is spherically symmetrical, and the other three, the $2p_x$, $2p_y$ and $2p_z$, lying along the x, y and z axes respectively, are dumb-bell shaped. For a hydrogen atom in its second excited state there are nine different orbitals which describe the possible electron distribution, the spherical 3s, the three dumb-bell shaped 3p and five 3d orbitals. For the third excited state there are 16 possible orbitals, a 4s, three 4p, five 4d and seven 4f.

The importance of the orbital picture is that the negative charge around a nucleus is imagined as 'smeared-out' rather than concentrated in tiny particles. The 'clouds' of negative charge can be used to explain such matters as the strength and direction of bonds in molecules, and the screening of the attractive force of the nucleus on an electron by other electrons in an atom.

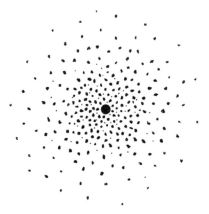

Fig. 10. Probability distribution
of 1s electron in the hydrogen
atom.

Orbital energies in light atoms other than hydrogen

A study of the atomic spectra of many elements led Pauli
to enunciate the exclusion principle; in modern terms it states
that no orbital can hold more than two electrons. In the boron
atom in its ground state the five electrons are distributed two
in the 1s orbital, two in the 2s orbital and one in a 2p orbital.
We say the electron configuration is $1s^2\ 2s^2\ 2p^1$. The orbital
energies in such an atom are illustrated in Fig. 11. No longer,
as in the H atom, are all the orbitals of principal quantum
number 2, i.e. the 2s and the 2p, of the same energy.

A 2s orbital is largely shielded from the coulombic
attraction of the nucleus by electrons in a 1s orbital but not
entirely, as a 2s electron spends some of its time close to the
nucleus. An electron in a 2p orbital is rather more strongly
shielded by 1s electrons. Similarly the 3s and 3p orbitals are
shielded to an increasing extent by electrons in the first and
second quantum levels. The order of orbital energies in an
atom other than hydrogen is always as shown in Fig. 11 although
the actual levels will depend, to some extent, on the total
number of electrons in orbitals closer to the nucleus. The
difference between energy levels are promotion energies
because they represent the energy needed to promote an electron from a low level to a higher one.

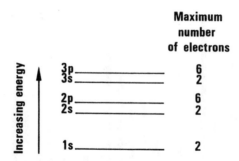

Fig. 11. Order of energy levels in an
atom containing several electrons.

The aufbau principle

It is possible to predict the electron configuration of any
atom in its ground state by adding Z electrons to the nucleus,
filling the available orbitals in ascending order of energy and
observing the exclusion principle.

Thus F $(Z = 9)$ has the electron configuration $1s^2 2s^2 2p^5$
Ne $(Z = 10)$ has the electron configuration $1s^2 2s^2 2p^6$
and Na $(Z = 11)$ has the electron configuration $1s^2 2s^2 2p^6 3s^1$

The principle on which this process is based, together with
rules regarding the electronic spins, which need not concern
us here, comprises the aufbau principle (German aufbau
= building up).

The first ionisation energies of elements 1–20

The variation of first ionisation energy, I_1, with atomic
number is shown for the elements from hydrogen to calcium in
Fig. 12. Two principal trends are apparent:

    (a) the tendency of I_1 to increase along a period (e.g. Li
       to Ne),

    (b) the tendency of I_1 to decrease down a group (e.g. He,
       Ne, Ar).

These tendencies can be explained by the use of the simple
concept of effective nuclear charge. The coulombic attraction
exerted by the nucleus of an atom on one of the outermost electrons is 'screened' by the other electrons, particularly by those

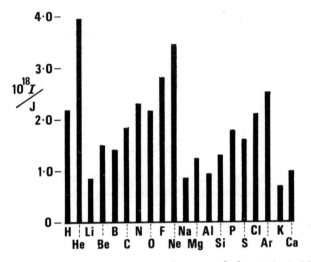

Fig. 12. First ionisation charges of elements 1-20.

in lower-energy orbitals, but to some extent even by those in the same orbital. Figure 13 shows the <u>screening constants</u> which represent the number of units of electronic charge by which a specified electron is shielded by another electron in a specified orbital. The figures were calculated by Slater from data on promotion energies and ionisation energies.

For example, the outermost electrons of the fluorine atom $(1s^2 2s^2 2p^5)$ are shielded from the nucleus by the following contributions from other electrons:

    6×0.35 electronic charges due to the other 2p and 2s electrons,

    2×0.85 electronic charges due to the 1s electrons.

The total screening, s, $= (6 \times 0.35) + (2 \times 0.85)$ electronic charges $= 3.8$. As $Z = 9$ for fluorine, the effective nuclear charge felt by the 2p electron under consideration is Z^*e,

$$\text{where } Z^*e = (Z - s)\,e$$
$$= (9 - 3.8)\,e = 5.2\,e$$

Thus the attractive force acting on the outermost electron is equivalent to that produced by $+5.2\,e$ at the nucleus. Using the same principle, Z^*e for a 2p electron in the neon atom is

FIG. 13

The Screening Effect of other Electrons on a Particular Electron in an Atom or Ion.

Electron considered	1s	2s or 2p	3s or 3p
			Others 0
Contribution of each		Others 0	$\begin{matrix}3p\ \text{——}\\3s\ \text{——}\end{matrix}\Big\}\ 0.35$
of the other elec- Others 0			
trons to its screening		$\begin{matrix}2p\ \text{——}\\2s\ \text{——}\end{matrix}\Big\}\ 0.35$	$\begin{matrix}2p\ \text{——}\\2s\ \text{——}\end{matrix}\Big\}\ 0.85$
from the nucleus. 1s —— 0.30		1s 0.85	1s —— 1.0

5.85 e, but for the sodium atom ($Z = 11$) the effective nuclear charge acting on the 3s electron is reduced to $\{11-2\,(1.0) -8\,(0.85)\}\,e = 2.2\ e$.

The calculation of theoretical ionisation energies using Slater's method requires certain assumptions, which are beyond the scope of this book, to be made about the energy levels of the orbitals. However, the figures already given are sufficient to show how general trends are correlated with Z^*. Furthermore Z^* is used in the calculation of electronegativity scales (p. 174).

Examples III

1. Calculate the frequencies of electromagnetic radiation of wavelength:

    (a) 15.3 nm (in the ultraviolet)
    (b) 584 nm (in the visible)
    (c) 48.7 μm (in the infrared)
    (d) 295 m (in the radiofrequency range)

2. Calculate the wave number of electromagnetic radiation of frequency:

    (a) 2.17×10^{16} Hz
    (b) 4.73×10^{14} Hz
    (c) 1.07×10^{6} Hz

3. For a particular metal the threshold frequency is 9.42×10^{14} Hz. Calculate the work function of the metal.

4. Calculate the kinetic energy of electrons emitted from the metal in Q.3 when light of wavelength 255 nm falls on its surface.

5. Calculate the energy of the photons in radiation of wavelength:

    (a) 635 nm (in the visible)
    (b) 18.7 nm (in the ultraviolet)
    (c) 58.6 μm (in the infrared)

6. Calculate the wavelength of a free electron after acceleration through a potential difference of:

    (a) 100 V
    (b) 10.0 V
    (c) 0.50 V

7. Calculate the wavelength of a proton (mass 1.67×10^{-27} kg) after acceleration through a potential of 50.0 V.

8. Calculate the wavelengths of the following lines in the hydrogen spectrum:

    (a) the fifth line in the Balmer series
    (b) the fifth line in the Lyman series
    (c) the fifth line in the Paschen series

9. Calculate the energy released when a hydrogen atom falls from the 4th excited state to the 2nd excited state.

10. Calculate the total screening, s, for:

    (a) the 3s electron in the sodium atom (electron configuration $1s^2\ 2s^2\ 2p^6\ 3s^1$)
    (b) a 2s electron in the carbon atom ($1s^2\ 2s^2\ 2p^2$)
    (c) A 2p electron in the F^- ion ($1s^2\ 2s^2\ 2p^6$)

11. Calculate the effective nuclear charge Z^*e acting on:

    (a) the 2s electron in the Be^+ ion
    (b) the 3s electron in the Mg^+ ion.

Chapter 4

The gas laws: the kinetic theory

Gases and vapours (p. 50) differ from solids and liquids in that they diffuse rapidly to fill a container and occupy volumes which are affected greatly by changes in temperature and pressure. Furthermore the volume changes are almost independent of the nature of the gas; this generality of behaviour led to the discovery of the gas laws.

Boyle's law

It was discovered by Boyle (1662) that for a given mass of gas at constant temperature the volume is inversely proportional to the pressure:

$$V \ \alpha \ 1/p$$

$$\text{or } pV \ = \ \text{a constant}$$

Example

A given mass of gas occupies 0.24 m^3 at a pressure of 1.27×10^5 N m^{-2}. Calculate the volume it would occupy at 7.60×10^4 N m^{-2} at the same temperature.

$$\text{Since } pV \ = \ \text{a constant}$$

$$p_2 V_2 \ = \ p_1 V_1$$

$$\text{and } \quad V_2 \ = \ \frac{p_1 V_1}{p_2}$$

Substituting values:

$$V_2 \ = \ \frac{1.27 \times 10^5 \text{ N m}^{-2} \times 0.24 \text{ m}^3}{7.6 \times 10^4 \text{ N m}^{-2}}$$

$$= \ 0.40 \text{ m}^3.$$

In early experiments it was found that nitrogen, hydrogen and oxygen obeyed the law almost perfectly, whereas carbon dioxide, sulphur dioxide and ammonia gave results which deviated from the law (p. 48).

Charles' law

Charles (1787) studied the effect of temperature on the volume of a gas at constant pressure. Expressed in modern terms, the volume of a given mass of gas at constant pressure is proportional to its thermodynamic temperature:

$$V \ \alpha \ T$$

$$\text{or } V \ = \ kT$$

Example

The volume of a given mass of gas is 0.36 m^3 at 288 K. At what temperature will the volume be 0.48 m^3 at the same pressure?

$$\text{Since } \frac{V}{T} \ = \ k$$

$$\frac{T_2}{V_2} \ = \ \frac{T_1}{V_1}$$

$$\text{and } \ T_2 \ = \ \frac{T_1 V_2}{V_1}$$

Substituting values:

$$T_2 \ = \ \frac{288 \text{ K} \times 0.48 \text{ m}^3}{0.36 \text{ m}^3}$$

$$= 384 \text{ K.}$$

Real gases show deviations from Charles' law as well as Boyle's law. The deviations decrease as the temperature is raised and the pressure is reduced. When a gas obeys both laws within the limits of the experimental accuracy it is said to behave as an ideal gas.

The equation of state for an ideal gas

The combination of Boyle's law and Charles' law produces the equation:

$$\frac{pV}{T} \ = \ \text{a constant for a given mass of gas.}$$

Example

A given mass of an ideal gas occupies 0.250 m^3 at 294 K and 1.41×10^5 N m^{-2}. At what pressure will the volume be 0.300 m^3 at 322 K?

Use $\dfrac{p_2 V_2}{T_2} = \dfrac{p_1 V_1}{T_1}$

i.e. $p_2 = \dfrac{p_1 V_1 T_2}{T_1 V_2}$

$= \dfrac{1.41 \times 10^5 \text{ N m}^{-2} \times 0.250 \text{ m}^3 \times 322 \text{ K}}{294 \text{ K} \times 0.300 \text{ m}^3}$.

$= 1.29 \times 10^5$ N m^{-2}.

The gas constant

The ideal gas equation can be written in the form

$$pV = nRT \qquad (20)$$

where n is the amount of gas and R is a universal constant, the gas constant, 8.314 J K^{-1} mol^{-1}.

Example

What amount of oxygen is contained in 0.010 m^3 of gas at a pressure of 1.01×10^5 N m^{-2} and a temperature of 300 K, assuming ideal behaviour?

Since $pV = nRT$

$n = \dfrac{pV}{RT}$

$= \dfrac{1.01 \times 10^5 \text{ N m}^{-2} \times 10^{-2} \text{ m}^3}{8.314 \text{ J K}^{-1} \text{ mol}^{-1} \times 300 \text{ K}}$

$= \dfrac{4.05 \text{ N m}}{\text{J mol}^{-1}}$

$= 4.05$ mol.

The Avogadro constant

Implicit in the ideal gas equation is the recognition that equal volumes of ideal gases, under the same conditions of temperature and pressure, contain equal amounts of gas, the word amount being defined as on p. 6. Avogadro (1811) introduced the concept of a gas molecule, the smallest particle capable of leading a separate existence in specified conditions, and postulated that equal volumes of different gases, under the same conditions, contained equal numbers of molecules. Experiment shows that one mole of a gas contains 6.023×10^{23} molecules; the quantity 6.023×10^{23} mol^{-1} is now known as the Avogadro constant, N_A.

Gaseous density and molar mass

The mass of a sample of gas divided by its volume is the gaseous density, ρ, at the specified temperature and pressure.

Example

At 273 K and 1.01×10^5 N m^{-2}, 4.46×10^{-4} kg of argon occupies 2.50×10^{-4} m^3. At this temperature and pressure

$$\rho = \frac{4.46 \times 10^{-4} \text{ kg}}{2.50 \times 10^{-4} \text{ m}^3} = 1.78 \text{ kg m}^{-3}$$

The molar mass, M, of a gas is the mass of one mole as defined on p. 6. An amount n of an ideal gas which occupies a volume V under specified conditions has a molar mass

$$M = \frac{\rho V}{n}$$

Since, for an ideal gas, $n = \dfrac{pV}{RT}$

$$M = \frac{\rho RT}{p} \qquad (21)$$

Thus, for argon, assuming the gas to be ideal,

$$M_{Ar} = \frac{1.78 \text{ kg m}^{-3} \times 8.31 \text{ J K}^{-1} \text{ mol}^{-1} \times 273 \text{ K}}{1.01 \times 10^5 \text{ N m}^{-2}}$$

$$= 4.00 \times 10^{-2} \text{ kg mol}^{-1} \qquad (J = N m)$$

Graham's law of diffusion

The phenomenon of diffusion can be simply shown in the laboratory by placing a few drops of bromine into a glass vessel; within seconds the vessel is filled with the reddish-brown vapour because the bromine molecules have diffused through the space. Gases will also diffuse through porous materials like unglazed earthenware, and this phenomenon is used to compare the rates at which different gases diffuse.

Graham (1829) observed that the rates at which gases diffuse under the same conditions of temperature and pressure are inversely proportional to the square roots of their densities. Thus for two gases of density ρ_1 and ρ_2 respectively, which diffuse at rates r_1 and r_2 respectively,

$$\frac{r_1}{r_2} = \sqrt{\frac{\rho_2}{\rho_1}}$$

if both gases are at the same temperature and pressure.

Since: $\rho = \frac{Mp}{RT}$ for an ideal gas, it follows that for constant values of p and T,

$$\rho \ \alpha \ M$$

and therefore

$$\frac{r_1}{r_2} = \sqrt{\frac{M_2}{M_1}} \tag{22}$$

Example

Under the same conditions of temperature and pressure, oxygen diffuses through a porous partition at the rate of 0.060 m^3 in 10 s, and chlorine at the rate of 0.100 m^3 in 25 s. If the molar mass of oxygen is 0.032 0 kg mol^{-1} calculate the molar mass of chlorine.

$$r_{1(O_2)} = \frac{0.060 \ m^3}{10 \ s} = 6.0 \times 10^{-3} \ m^3 \ s^{-1}$$

$$r_{2(Cl_2)} = \frac{0.100 \ m^3}{25 \ s} = 4.0 \times 10^{-3} \ m^3 \ s^{-1}$$

Since $\dfrac{r_1}{r_2} = \sqrt{\dfrac{M_2}{M_1}}$

$$\frac{M_2}{M_1} = \frac{r_1^2}{r_2^2}$$

and $M_2 = \dfrac{M_1 r_1^2}{r_2^2}$

$$= \frac{3.20 \times 10^{-2} \ kg \ mol^{-1} \times (6.0 \times 10^{-3} \ m^3 \ s^{-1})^2}{(4.0 \times 10^{-3} \ m^3 \ s^{-1})^2}$$

$$= 7.20 \times 10^{-2} \ kg \ mol^{-1}.$$

Dalton's law of partial pressures

The total pressure of a mixture of gases is the sum of the partial pressures of the constituent gases (Dalton, 1801):

$$p = p_A + p_B + p_C \qquad (23)$$

where p_A, p_B and p_C are partial pressures of the constituents A, B and C respectively. The partial pressure is defined as the pressure each gas would exert if it alone occupied the vessel at that temperature. In a mixture of ideal gases

$$p_A = \frac{n_A}{n} \times p \qquad (24)$$

where n_A is the amount of A and n is the total amount of gas.

Example

A globe of volume 2.00×10^{-4} m^3 containing oxygen at a pressure of 1.00×10^5 N m^{-2} was put in communication with a globe of volume 4.00×10^{-4} m^3 containing nitrogen at a pressure of 0.25×10^5 N m^{-2} and the gases were allowed to mix. Calculate (a) the partial pressure of oxygen (b) the total pressure, assuming that there was no change in temperature.

For oxygen:

$$p_{O_2} \ (= p_2) = \frac{p_1 V_1}{V_2}$$

$$= \frac{1.00 \times 10^5 \text{ N m}^{-2} \times 2.00 \times 10^{-4} \text{ m}^3}{6.00 \times 10^{-4} \text{ m}^3}$$

$$= 3.33 \times 10^4 \text{ N m}^{-2}$$

Similarly for nitrogen:

$$p_{N_2} = \frac{0.25 \times 10^5 \text{ N m}^{-2} \times 4.00 \times 10^{-4} \text{ m}^3}{6.00 \times 10^{-4} \text{ m}^3}$$

$$= 1.67 \times 10^4 \text{ N m}^{-2}$$

Thus $p = p_{O_2} + p_{N_2} = (3.33 + 1.67) \times 10^4$ N m^{-2}

$$= 5.00 \times 10^4 \text{ N m}^{-2}$$

The kinetic theory of gases

The kinetic theory, which was developed in the years 1860–65 to explain the behaviour of gases, is based on the idea that a gas consists of perfectly elastic, spherical molecules which are in continuous, high-speed, random motion. The pressure exerted by the gas is due to the bombardment of the walls of the containing vessel by the molecules which, except at very high pressures, are separated by distances far greater than their own diameters. The effect of raising the temperature of the vessel is to increase the kinetic energies of the molecules.

The relation between the pressure of a gas and the kinetic energy of its molecules

Imagine a cube of side l containing x molecules with different speeds c_1, c_2, c_3 etc. Figure 14 represents a single molecule with a velocity c_1 in the direction shown. Its velocity can be resolved in three directions parallel to the edges of the cube, and

$$c_1^2 = c_x^2 + c_y^2 + c_z^2.$$

Consider a molecule of a mass m whose velocity relative to the x-axis is c_x and its momentum mc_x. Assuming it to be perfectly elastic, its change of momentum relative to the x-axis on striking the wall A is $mc_x - (-mc_x) = 2\ mc_x$. The time taken for it to rebound from the opposite wall, where the resolved momentum again alters by $2\ mc_x$, and hit wall A again is $\frac{2\ l}{c_x}$. Thus the rate of change of momentum relative to the x-axis is $(2\ mc_x + 2\ mc_x)\frac{c_x}{2\ l}$

$$= \frac{2\ mc_x^2}{l}$$

Similarly the rates of change of momentum relative to the y- and z-axes are $\frac{2\ mc_y^2}{l}$ and $\frac{2\ mc_z^2}{l}$, and the total rate of change of momentum is

$$\frac{2\ mc_x^2}{l} + \frac{2\ mc_y^2}{l} + \frac{2\ mc_z^2}{l} = \frac{2\ mc_1^2}{l}$$

For all the molecules the rate of change of momentum is

$$\frac{2\ m}{l}(c_1^2 + c_2^2 + c_3^2 + \ -\ -\ -\ -\ -)$$

We define the root mean square velocity, $\bar{c}$, as the square root of $\bar{c}^2$ which is the mean value of c_1^2, c_2^2, c_3^2 etc.

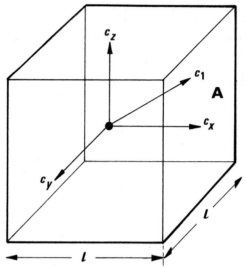

Fig. 14. Resolution of velocity c along
x, y, z axes.

Since total force = rate of change of momentum

$$= \frac{2\ xm\bar{c}^2}{l},$$

and this force acts on six sides of the cube each of area l^2, the pressure exerted by the bombarding molecules on each surface is

$$p \;=\; \frac{2\ xm\bar{c}^2}{l} \Big/ 6\,l^2 \;=\; \frac{xm\bar{c}^2}{3\ l^3} \;=\; \frac{xm\bar{c}^2}{3\ V}$$

Hence $\qquad\qquad pV \;=\; \dfrac{xm\bar{c}^2}{3}$ $\qquad\qquad$ (25)

Although this equation has been derived using a cube as the basic shape, any shape could be used by assuming it to be made of a number of small cubes.

Equation 25 can be used to deduce the laws relating to ideal gases.

(a) Since the mean kinetic energy of the gas molecules, $\frac{1}{2}\ m\bar{c}^2$, is a constant for a given temperature:

$$pV \;=\; \frac{xm\bar{c}^2}{3} \;=\; \text{a constant (Boyle's law)}$$

(b) Since the kinetic energy $\frac{1}{2} m\bar{c}^2$ = a constant $\times T$,

$$pV = \text{a constant} \times T$$

and $\qquad V = kT$ at constant pressure (Charles' law).

(c) Consider equal volumes of two gases, A and B, at equal pressures and at the same temperature, mixed together without any reaction.

Since $p_A = p_B$,

$$V_A = V_B,$$

and $\quad T_A = T_B$ (i.e. $\frac{1}{2} m_A \bar{c}_A^2 = \frac{1}{2} m_B \bar{c}_B^2$)

$$\therefore x_A = x_B \quad \text{(Avogadro's law)}.$$

(d) Since $\bar{c}^2 = \dfrac{3 pV}{xm}$ (from Eqn. 25)

and $\qquad \rho = \dfrac{xm}{V}$

$$\therefore \bar{c}^2 = \dfrac{3p}{\rho}$$

and $\qquad \bar{c} = \sqrt{\dfrac{3p}{\rho}}$ \hfill (26)

Thus for two different gases at the same temperature and pressure

$$\frac{\bar{c}_A}{\bar{c}_B} = \sqrt{\frac{\rho_B}{\rho_A}} \, ,$$

and since rate of diffusion is proportional to molecular speed,

$$\frac{r_A}{r_B} = \sqrt{\frac{\rho_B}{\rho_A}} \quad \text{(Graham's law)}$$

(e) If p_A is the pressure exerted by the molecules of gas A in a volume V and p_B is the pressure exerted by gas B

$$p_A = \frac{x_A m_A \bar{c}_A^2}{3V}$$

$$p_B = \frac{x_B m_B \bar{c}_B^2}{3V}$$

Then provided the molecules do not interact with one another

$$p = \frac{x_A m_A \bar{c}_A^2 + x_B m_B \bar{c}_B^2}{3V} = p_A + p_B \quad \text{(Dalton's law)}$$

Example

At 273 K and 1.01×10^5 N m^{-2}, the density of oxygen is
1.43 kg m^{-3}. Calculate the root mean square velocity of the
molecules under these conditions.

$$\bar{c} = \sqrt{\frac{3p}{\rho}}$$

$$= \sqrt{\frac{3 \times 1.01 \times 10^5 \text{ N m}^{-2}}{1.43 \text{ kg m}^{-3}}}$$

$$= \sqrt{2.12 \times 10^5 \text{ m}^2 \text{ s}^{-2}} \quad (\text{N} = \text{kg m s}^{-2})$$

$$= 460 \text{ m s}^{-1}$$

The behaviour of real gases

Experiments on the behaviour of gases at high pressure
show that real gases are less compressible than an ideal gas
would be; on the other hand, at low pressures real gases are
more easily compressed than expected from the ideal gas
equation. The deviations from ideal behaviour have been attri-
buted to two causes, the existence of attractive forces between
the molecules and the fact that the sizes of the molecules are
not entirely negligible as assumed in the simple kinetic theory.

The attractive forces between molecules are electrostatic in nature. Even in a helium molecule, which consists
of a nucleus and two electrons, the centre of negative charge
is not necessarily at the same point as the centre of positive
charge because the electron density at any instant is not spherically symmetrical. The momentary polar character of the
molecule is sufficient to induce polarity in surrounding molecules and cause electrostatic attraction; this is known as the
dispersion interaction effect and it occurs in all gases, the
size of the effect increasing with the size and complexity of the
molecules. In gases such as HCl in which the molecules are
permanent dipoles (p. 172) there is additional attraction
between molecules due to the so-called orientation interaction.

The van der Waals equation

Various attempts have been made to develop equations of
state which will apply to real gases. Of the scores of equations

which have been suggested, we shall use one of the simplest,
due to van der Waals (1873), which represents a fairly realistic
description of the behaviour of most gases unless their densities are very high. For one mole of gas   ⟵

$$(p + \frac{a}{V^2})\ (V - b)\ =\ RT. \tag{27}$$

The term $\frac{a}{V^2}$ which is added to the pressure is the correction

for intermolecular attraction, the term b which is subtracted
from the volume is called the covolume; the factor $(V - b)$ ⟵
represents the effective compressible volume. The value of b
is generally taken to be four times the volume of the molecules
considered as spheres.

Example

Calculate the pressure exerted by 0.100 mole of SO_2 which
occupies 1.00×10^{-3} m^3 at 373 K (a) using the ideal gas equation (b) using the van der Waals equation with $a = 0.680$ N m^4
mol^{-2} and $b = 5.64 \times 10^{-5}$ m^3 mol^{-1}.

(a) $pV = nRT$

$$\therefore p = \frac{0.100 \text{ mol} \times 8.314 \text{ J K}^{-1} \text{ mol}^{-1} \times 373 \text{ K}}{1.00 \times 10^{-3} \text{ m}^3}$$

$$= 3.10 \times 10^5 \text{ N m}^{-2}.$$

(b) If 0.100 mol occupies 1.00×10^{-3} m^3 under the conditions, 1.00 mol occupies 1.00×10^{-2} m^3, i.e. $V = 1.00 \times 10^{-2}$
m^3 mol^{-1}. Now substituting in the van der Waals' equation,

$$p = \frac{RT}{V - b} - \frac{a}{V^2} \text{ (rearranged form of Eqn. 27)}$$

$$= \frac{8.314 \text{ J K}^{-1} \text{ mol}^{-1} \times 373 \text{ K}}{(1.00 \times 10^{-2} - 5.64 \times 10^{-5}) \text{ m}^3 \text{ mol}^{-1}} - \frac{0.680 \text{ N m}^4 \text{ mol}^{-2}}{(1.00 \times 10^{-2} \text{ m}^3 \text{ mol}^{-1})^2}$$

$$= 3.05 \times 10^5 \text{ N m}^{-2}.$$

Table X gives some values of the van der Waals constants a
and b. It will be seen that a is least for small, non-polar
molecules and b is least when the molecules contain few atoms.

Critical phenomena

Figure 15 shows the experimental 'pV isothermals'
obtained for carbon dioxide by Andrews (1869). Above 304.2 K

TABLE X
Van der Waals' Constants for Some Gases

	$a/\text{N m}^4 \text{ mol}^{-2}$	$10^5 b/\text{m}^3 \text{ mol}^{-1}$
He	0.00344	2.37
H_2	0.0247	2.66
N_2	0.141	3.91
NO_2	0.535	4.42
C_2H_6	0.555	6.38
SO_2	0.680	5.64

the volume varies smoothly with pressure, and the gas does not liquefy no matter how high the pressure applied. Below 304.2 K, the gas, now correctly called a vapour, is liquefied, with a concomitant sharp drop in volume, at a pressure which depends upon the temperature. The curve for 304.2 K shows a point of inflexion at $7.3 \times 10^6 \text{ N m}^{-2}$. The temperature 304.2 K is the critical temperature for carbon dioxide. The critical temperature, T_C, of a gas is the temperature above which it cannot be liquefied no matter how high the pressure. The critical pressure, p_C, is the lowest pressure which will liquefy the gas at its critical temperature. It is also useful to define a critical molar volume, V_C, the volume of one mole at T_C and p_C.

It can be seen in Table XI that H_2, N_2 and O_2 have to be cooled appreciably below ordinary atmospheric temperature before the gases can possibly be liquefied.

The relation between the critical constants and the van der Waals constants

Eqn. (26) can be rearranged in the form:

$$pV - pb + \frac{a}{V} - \frac{ab}{V^2} = RT$$

Multiplying by $\frac{V^2}{p}$:

$$V^3 - bV^2 + \frac{aV}{p} - \frac{ab}{p} = \frac{RTV^2}{p}$$

Rearranging:

$$V^3 - \left(\frac{RT+b}{p}\right) V^2 + \frac{a}{p}V - \frac{a}{p}b = 0 \qquad (28)$$

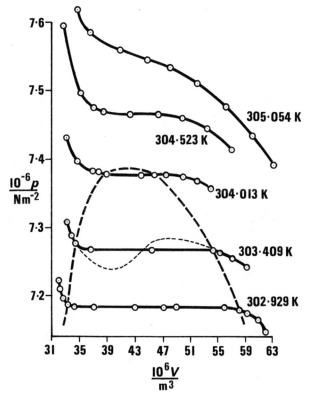

Fig. 15. Andrews' curves for CO_2.

At the critical point $V = V_c$, i.e. $(V - V_c)^3 = 0$

or $V^3 - 3 V_c V^2 + 3 V_c^2 V - V_c^3 = 0$

Comparing the coefficients with those of Eqn. (28),

$$3 V_c = \frac{RT_c + b}{p} \qquad \text{(i)}$$

$$3 V_c^2 = \frac{a}{p_c} \qquad \text{(ii)}$$

$$\text{and} \quad V_c^3 = \frac{ab}{p_c} \qquad \text{(iii)}$$

Dividing (iii) by (ii), $V_c = 3\,b$ (29)

Substituting in (ii), $p_c = \dfrac{a}{27\,b^2}$ (30)

Substituting in (i), $T_c = \dfrac{8\,a}{27\,Rb}$ (31)

TABLE XI

Some Values of T_c and p_c

	T_c/K	$10^{-6}p_c/\text{N m}^{-2}$
H_2	38	2.0
N_2	127	3.3
O_2	155	5.0
CO_2	304	7.3
Cl_2	417	7.6
SO_2	430	7.6

Example

For Cl_2, $T_c = 417$ K and $p_c = 7.6 \times 10^6$ N m^{-2}. Calculate the van der Waals constants a and b.

$$\frac{p_c}{T_c} = \frac{a}{27\, b^2} \times \frac{27\, Rb}{8\, a} = \frac{R}{8\, b}$$

$$\text{Thus } b = \frac{T_c \times R}{8\, p_c} = \frac{417 \text{ K} \times 8.314 \text{ J K}^{-1} \text{ mol}^{-1}}{8 \times 7.6 \times 10^6 \text{ N m}^{-2}}$$

$$= 5.70 \times 10^{-5} \text{ m}^3 \text{ mol}^{-1} \qquad (\text{J} = \text{N m})$$

Since $\dfrac{8\, a}{27\, Rb} = T_c$,

$$a = \frac{27\, RbT_c}{8}$$

$$= \frac{27 \times 8.314 \text{ J K}^{-1} \text{ mol}^{-1} \times 5.70 \times 10^{-5} \text{ m}^3 \text{ mol}^{-1} \times 417 \text{ K}}{8}$$

$$= 0.667 \text{ N m}^4 \text{ mol}^{-2} \qquad\qquad (\text{J} = \text{N m})$$

Examples IV

1. A certain mass of an ideal gas has a volume of 3.00×10^{-3} m^3 at a pressure of 1.01×10^5 N m^{-2}. Calculate its volume at a pressure of 7.1×10^4 N m^{-2} at the same temperature.

2. A given amount of an ideal gas has a volume of 4.00×10^{-3} m^3 at a pressure of 1.01×10^4 N m^{-2}. Calculate the pressure required to compress it to a volume of 2.50×10^{-4} m^3 at the same temperature.

3. An ideal gas has a volume of 3.50×10^{-4} m^3 at 300 K. Calculate the volume occupied by the gas at 450 K at the same pressure.

4. Calculate the temperature to which a gas must be raised to expand to fill a volume of 3.20×10^{-4} m^3 if its volume is 1.80×10^{-4} m^3 at 280 K and the pressure remains constant.

5. A given mass of an ideal gas occupies 2.80×10^{-4} m^3 at 8.50×10^4 N m^{-2} and 300 K. Calculate its volume at 1.01×10^5 N m^{-2} and 273 K.

6. An amount of an ideal gas occupies 4.20×10^{-4} m^3 at 9.80×10^4 N m^{-2} and 290 K. Calculate the temperature at which it occupies 3.90×10^{-4} m^3 at 1.01×10^5 N m^{-2}.

7. Calculate the volume occupied by 0.150 moles of an ideal gas at 1.01×10^5 N m^{-2} and 295 K.

8. Calculate the amount of ideal gas which occupies 2.20×10^{-3} m^3 at 5.80×10^4 N m^{-2} and 300 K.

9. The density of krypton is 3.44 kg m^{-3} at 298 K and 1.01×10^5 N m^{-2}. Calculate its molar mass, assuming it to behave ideally under the above conditions.

10. The density of methane at 298 K and 1.06×10^5 N m^{-2} is 6.90×10^{-1} kg m^{-3}. Assuming it to behave ideally, calculate its molar mass.

11. In 30 s, 0.180 m^3 of oxygen diffused through a porous plate. How long will it take 0.400 m^3 of CO_2 to diffuse through the plate at the same temperature and pressure? $M(O_2) = 0.032$ kg mol^{-1} and $M(CO_2) = 0.044$ kg mol^{-1}.

12. A volume of 0.125 m^3 of an ideal gas A, measured at 6.06×10^4 N m^{-2}, and 0.150 m^3 of gas B at 8.08×10^4 N m^{-2} are passed into a vessel having a capacity of 0.500 m^3. Calculate (a) the partial pressures of A and B, (b) the total pressure of gas in the vessel, the temperature being constant.

13. The density of hydrogen at 273 K and 1.01×10^5 N m^{-2} is 8.96×10^{-2} kg m^{-3}. Calculate the root mean square velocity of the hydrogen molecules under these conditions.

14. The density of ammonia at 300 K and 1.02×10^5 N m^{-2} is 7.00×10^{-1} kg m^{-3}. Calculate the root mean square velocity of ammonia molecules under these conditions.

15. Use Table X and the van der Waals equation to calculate (a) the pressure exerted by 1.00 mole of C_2H_6 which occupies 4.40 $\times$ 10^{-4} m^3 at 300 K. Calculate (b) the pressure which the ethane would exert if it behaved ideally.

16. Use Table X and the van der Waals equation to calculate (a) the pressure exerted by 0.150 mole N_2 at 300 K which occupies 4.50 $\times$ 10^{-4} m^3. Calculate also (b) the pressure which the nitrogen would exert if it behaved ideally.

17. Use Table X to calculate the approximate radius of the helium molecule, the so-called van der Waals radius, assuming b to represent four times the volume of one mole of helium.

18. Use Table X to calculate T_c, p_c and V_c for C_2H_6.

19. Use Table X to calculate T_c, p_c and V_c for helium.

20. Use Table XI to calculate the van der Waals constants for oxygen.

Chapter 5

Ionic crystals

X-ray analysis of crystals: the Bragg equation

In 1912, W. L. Bragg investigated the reflection of monochromatic (i.e. uniform wavelength) X-ray beams from the
surfaces of crystals such as sodium chloride and other simple
salts. He found that, for a particular salt, there were certain
angles between the incident beam and the surface which gave
rise to strong reflection. Figure 16 illustrates his explanation;
the lines AB, A'B' and A"B" represent parallel rays reflected
from successive planes in a crystal. The path difference
between rays AB and A'B', reflected from planes separated by
a distance d is ED = $2d \sin \theta$, since B'B = B'D and BD = $2d$.
Reinforcement of reflection can occur only if the electromagnetic waves along BC are in phase, that is if the difference in
path length represents an integral number of wavelengths.

Thus the condition for reinforcement of reflection is

$$n = 2\,d\,\sin\,\theta \quad \text{(the Bragg equation)} \tag{32}$$

The intensity of the reflected beam falls sharply to zero when
the glancing angle is altered slightly from that required for
strong reflection, because the reflected waves are out of phase.

The method was used to find the distances between layers
of atomic nuclei, and later developed to determine the positions
of nuclei relative to one another in space.

Example

A narrow beam of X-rays of wavelength λ = 58.5 pm was
found to give reflection maxima from the face of a sodium
chloride crystal when the glancing angles between the beam and

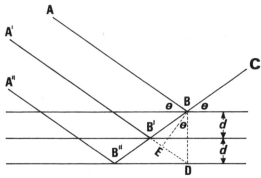

Fig. 16.   Illustrating the derivation of the
Bragg equation.

the face were 5.90°, 11.85° and 17.95°. The sines of these
angles are respectively 0.1028, 0.2054 and 0.3080; notice that
they are in the ratio 1:2:3 depending upon whether $n = 1$, 2 or
3; these are called the reflections of first, second and third
order. Thus, substituting in the Bragg equation for the first-
order reflection, d is given by

$$d = \frac{\lambda}{2 \times 0.1028} = \frac{58.5 \text{ pm}}{0.2056}$$

$$= 284 \text{ pm}$$

Thus the distance between successive planes of atomic nuclei
parallel to the reflecting face in sodium chloride is found to be
284 pm.

Ionic lattices

The X-ray analysis of crystal structure shows that solids
like sodium chloride consist of regular three-dimensional
assemblages of cations and anions. Figure 17 represents the
sodium chloride lattice. Each Na^+ ion is surrounded by six
Cl^- ions as its nearest neighbours, and each Cl^- ion by six
Na^+ ions. Both the Na^+ ion and the Cl^- ion in the structure
have the co-ordination number 6; the structure exemplifies 6:6
co-ordination. This is a particularly common form of ionic
lattice; most of the alkali-metal halides and many oxides of
dipositive metals (e.g. CaO and MgO) are examples.

Although most alkali-metal halides have the NaCl structure, the crystals of CsCl, CsBr and CsI have the caesium
chloride lattice which exemplifies 8:8 co-ordination (Fig. 18).

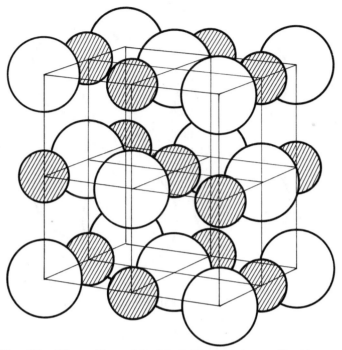

Fig. 17. The sodium chloride lattice. 6:6 coordination.
(Na$^+$ shaded.)

Every Cs$^+$ion is surrounded in a regular three-dimensional
pattern by eight Cl$^-$ ions as its nearest neighbours and every
Cl$^-$ ion by eight Cs$^+$ ions. Again, although MgO and CaO have
the NaCl structure ZnO is an example of the wurtzite lattice
(Fig. 19) with 4:4 co-ordination.

Examples of 5:5 co-ordination, 7:7 co-ordination or $n{:}n$
co-ordination with $n > 8$ are not known in ionic crystals,
presumably because they cannot produce regular three-dimensional patterns.

Lattice energies

In order to understand why the Na$^+$ ion in NaCl is six-
co-ordinate whereas the Cs$^+$ ion in CsCl is eight-co-ordinate
and the Zn^{2+} ion in ZnO is four-co-ordinate, we must first
consider the energy changes which accompany the formation of
an ionic lattice from its individual ions.

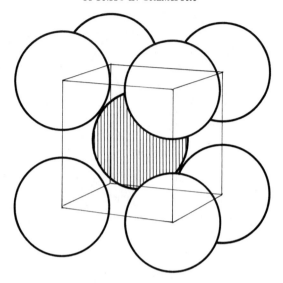

Fig. 18. The caesium chloride lattice.
8:8 coordination.

The force of attraction between a point charge $+e$ and a
point charge $-e$ separated by a distance r in a vacuum is

$$F = \frac{e^2}{4\pi\epsilon_0 r^2} \quad \text{(p. 4)} \tag{33}$$

The energy released when two such charges, approaching from
infinity, reach a distance r_0 from one another is

$$E = \int_{\infty}^{r_0} \frac{e^2}{4\pi\epsilon_0 r^2}\ \mathrm{d}r = \frac{e^2}{4\pi\epsilon_0 r_0} \tag{34}$$

For two such point charges approaching from infinite distance
to $r_0 = 284$ pm, the distance between ionic centres in the NaCl
crystal, the energy release is therefore

$$E = \frac{(1.602 \times 10^{-19}\ \text{A s})^2}{4 \times 3.142 \times 8.854 \times 10^{-12}\ \text{kg}^{-1}\ \text{m}^{-3}\ \text{s}^4\ \text{A}^2 \times 2.84 \times 10^{-10}\ \text{m}}$$

$$= 8.12 \times 10^{-19}\ \text{kg m}^2\ \text{s}^{-2}$$

$$= 8.12 \times 10^{-19}\ \text{J}.$$

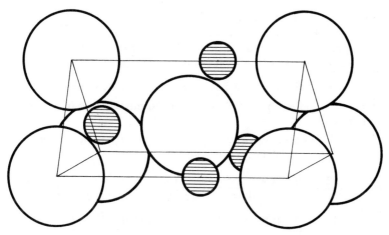

Fig. 19. The wurtzite lattice. 4:4 coordination (cations shaded).

But the energy release when 6.02×10^{23} Na$^+$ ions and the same number of Cl$^-$ ions approach from infinity to form solid NaCl is not simply

$$8.12 \times 10^{-19} \text{ J} \times 6.02 \times 10^{23} \text{ mol}^{-1}$$

$$= 489 \text{ kJ mol}^{-1}.$$

There are two reasons for this.

(a) The ions are not point charges, but consist of positive nuclei surrounded by electronic clouds which interact with neighbouring electronic clouds.

(b) The electric field round an ion is due not only to the ions of opposite charge adjacent to it, but also to ions further from it.

Mathematical refinement of the equation for the total energy release, taking the above factors into consideration, has led to the development of equations for a more realistic calculation of the lattice energy of the ionic solid; this is the energy released when the gaseous ions, originally at infinite distances apart, arrange themselves in their equilibrium positions in the crystal lattice. The equation we shall use is that due to Born and Landé:

$$U = \frac{N_A A_a(z_+)(z_-)e^2}{4\pi\epsilon_0 r_0}(1-n) \tag{35}$$

in which A_a is a constant for the type of lattice, the Madelung constant calculated from a consideration of (b) above; $+z_+e$ and $-z_-e$ are the charges on cation and anion respectively; n, the Born exponent, is calculated from a consideration of (a) above.

For a solid like NaCl, containing an Ne type ion (Na^+) and an Ar type ion (Cl^-) (see Table XIII) the mean value of 8 ($= \frac{7+9}{2}$) is taken for n.

Using the Born-Landé equation, the lattice energy of sodium chloride is therefore given by

$$U = \frac{6.02 \times 10^{23} \text{ mol}^{-1} \times 1.748 \times (1.602 \times 10^{-19} \text{ A s})^2 \times (1 - \frac{1}{8})}{4 \times 3.142 \times 8.854 \times 10^{-12} \text{ kg}^{-1} \text{m}^{-3} \text{s}^4 \text{ A}^2 \times 2.84 \times 10^{-10} \text{ m}}$$

$$= 7.48 \times 10^5 \text{ kg m}^2 \text{ s}^{-2} \text{ mol}^{-1} = 748 \text{ kJ mol}^{-1}$$

since e (the charge on an electron) $= 1.602 \times 10^{-19}$ A s,
   ϵ_0 (the permittivity of a vacuum) $= 8.854 \times 10^{-12} \text{ kg}^{-1} \text{m}^{-3} \text{s}^4 \text{A}^2$,
   N_A (the Avogadro constant) $= 6.02 \times 10^{23} \text{ mol}^{-1}$,
   n (the Born exponent) $= 8$ (see Table XIII)
and A_a (the Madelung constant) $= 1.748$ (see Table XII).

There are two factors which influence the lattice energy of an ionic solid which have direct bearing on its structure.

(i) The value of U will be a maximum if r_0 is as short as possible — the anions and cations actually 'touch' one another.

(ii) The value of U increases with increase in the co-ordination numbers of the ions. The reader will notice that A_a (wurtzite) $< A_a$ (NaCl) $< A_a$ (CsCl) (Table XII).

The radius ratio rule

The preferred co-ordination number of an ion can be predicted by considering ions to be hard spheres. A Cs^+ ion in CsCl, considered as a hard sphere, is big enough to hold apart eight Cl^- ions, but an Na^+ ion in the same environment would 'rattle about' inside the arrangement of eight Cl^- ions, and condition (i) above would not be fulfilled. Coulombic energy would be released if two of the anions were pushed out by the others so that the cation was in contact with all the anions surrounding it. The ratio cationic radius:anionic radius, r_c/r_a, is therefore an important factor in determining co-ordination number.

Figure 20 represents the smallest cation capable of holding apart four anions in the same plane. Its radius r_c $= \sqrt{2}\, r_a - r_a$, and $\frac{r_c}{r_a} = \sqrt{2} - 1 = 0.414$. This is the smallest

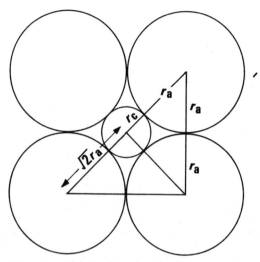

Fig. 20. Illustrating the radius ratio rule.

radius ratio r_c/r_a for which 6-co-ordination of the cation is
possible (in addition to the four in the plane, one anion can be
accommodated above, and one below the cation). For lower
values of r_c/r_a the cation will be 4-co-ordinate; for much
higher values, of course, 8-co-ordination will become possible
(Q. 8, p. 67).

The prediction of co-ordination number using the radius
ratio rule occasionally breaks down, because ions are not hard
spheres. There is always some overlapping of the electronic
clouds of adjacent ions (covalent character) the extent of which
depends on the character of the ions. Useful general rules are
that the tendency to covalent character in an ionic crystal will
increase (a) as r_c/r_a decreases, (b) as the electrostatic
charge on the cation increases.

Ionic lattices containing unequal numbers of cations and anions

Figure 21 illustrates the CaF_2 (fluorite) lattice. The
radius ratio $r_{Ca^{2+}}/r_{F^-}$ is similar to that in CsCl, and the Ca^{2+}
ion is 8-co-ordinate. But the F^- ion is only 4-co-ordinate as
there are only half as many Ca^{2+} ions as F^- ions. This is 8:4
co-ordination. In a similar way AlF_3 (Fig. 22) has what is
effectively an NaCl lattice with three-quarters of the cations
and one-quarter of the anions missing – 6:2 co-ordination
– and Al_2O_3 (corundum) in Fig. 23 has an NaCl lattice with
one-third of the cations missing – 6:4 co-ordination.

TABLE XII

Some Madelung constants

	A_a
NaCl lattice (p. 57)	1.748
CsCl lattice (p. 58)	1.763
Wurtzite lattice (p. 59)	1.641
Fluorite lattice (p. 64)	2.519

Ionic radii

The X-ray analysis of a crystal enables $r_a + r_c$, the internuclear distance, to be measured, but the individual ionic radii must be calculated. The method due to Pauling uses the concept of effective nuclear charge (p. 35). To obtain the radii of two singly-charged, isoelectronic ions like Na^+ and F^- (both $1s^2\, 2s^2\, 2p^6$) he divided the internuclear distance in inverse ratio of the effective nuclear charge exerted at the outer electronic shell of each ion. For a $1s^2\, 2s^2\, 2p^6$ ion $Z^* = Z - 8\,(0.35) -2\,(0.85) = Z - 4.5$.

Thus for Na^+, $Z^* = 11 - 4.5 = 6.5$
and for $\quad F^-$, $Z^* = 9 - 4.5 = 4.5$.

The internuclear distance, here 231 pm, is divided in the inverse ratio of these charges:

$$\frac{r_{Na^+}}{r_{F^-}} = \frac{4.5}{6.5}$$

Thus $r_{F^-} = \dfrac{6.5}{6.5 + 4.5} \times 231 \text{ pm} = 136 \text{ pm}$

and $r_{Na^+} = (231 - 136)\,\text{pm} = 95 \text{ pm}$

Pauling has also devised methods for calculating the radii of ions of charge greater than one and for estimating the effect of change of co-ordination number on radius, but these do not concern us here.

The Avogadro constant

In KCl, the distance $r_{K^+} + r_{Cl^-} = 314$ pm by X-ray analysis. Reference to Fig. 17 indicates that a cube of side 628 pm contains 4 K^+ and 4 Cl^- ions. One mole of KCl weighs 7.46×10^{-2} kg and the density of the crystal is 2004 kg m^{-3}.

Thus one mole occupies $\dfrac{7.46 \times 10^{-2} \text{ kg}}{2004 \text{ kg m}^{-3}} = 3.72 \times 10^{-5} \text{ m}^3$.

Born Exponents for Different Ion Types

	n
He ($1s^2$)	5
Ne ($2s^2 2p^6$)	7
Ar ($3s^2 3p^6$) and Cu^+ ($3d^{10}$)	9
Kr ($4s^2 4p^6$) and Ag^+ ($4d^{10}$)	10
Xe ($5s^2 5p^6$) and Au^+ ($5d^{10}$)	12

But one pair of ions occupies $\dfrac{(6.28 \times 10^{-10} \text{ m})^3}{4}$.

Thus one mole contains $\dfrac{3.72 \times 10^{-5} \text{ m}^3 \times 4}{(6.28 \times 10^{-10} \text{ m})^3}$ pairs of ions = 6.01 $\times 10^{23}$ pairs of ions.

The accepted value of N_A is 6.02×10^{23} mol^{-1}.

In fact X-ray analysis enables the distances between crystal planes to be measured with four-figure precision, and as densities can also be measured accurately, the above principle is one of the best for the determination of the Avogadro constant.

The thermochemistry of crystal formation

The importance of the lattice energy in the formation of a crystal can be illustrated by considering the formation of magnesium oxide from its elements, a strongly exothermic reaction:

$$\text{Mg (s)} + \tfrac{1}{2}\text{O}_2 \text{ (g)} = \text{MgO (s)} \quad \Delta H^\circ = -603 \text{ kJ mol}^{-1}. \quad \text{(a)}$$

(The reader should consult Chapter 6 for definitions of thermochemical terms).

The processes which lead to the formation of Mg^{2+} ions and O^{2-} ions from the elements are, however, endothermic:

$\Delta H^\circ / \text{kJ mol}^{-1}$

Atomisation of metal:	Mg (s) = Mg (g)	+ 146
Ionisation of metal atoms:	Mg (g) = Mg^{2+} + 2e	+2178
Dissociation of O_2 molecules:	$\tfrac{1}{2}O_2$ (g) = O (g)	+ 249
Formation of O^{2-} ions from atoms:	O (g) + 2e = O^{2-} (g)	+ 739
Formation of gaseous ions from elements (by addition):	Mg (s) + $\tfrac{1}{2}O_2$ (g) = Mg^{2+} (g) + O^{2-} (g)	+3312 (b)

Fig. 21. The CaF_2 (fluorite) lattice: 8:4
co-ordination (adjacent layers). Ca^{2+} ions
are shaded. In the corresponding layers
of the CsCl lattice (Fig. 18) the cations
are spaced as closely as the anions.

The lattice energy of MgO is very large, $U = 3915$ kJ mol^{-1},
i.e. for $Mg^{2+}(g) + O^{2-}(g) = MgO(s)$ $\Delta H^\circ = -3915$ kJ mol^{-1}. (c).

Adding (c) and (b) gives (a) above. Clearly the energy
released in the formation of the crystal lattice is an extremely
important factor in the thermodynamic stabilisation of the solid.

The Born-Haber cycle

Different paths to the formation of a crystal MX from its
elements are shown in the Born–Haber cycle below:

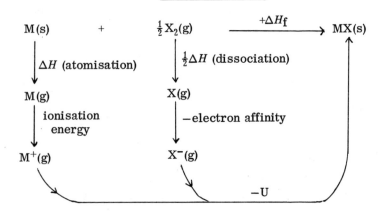

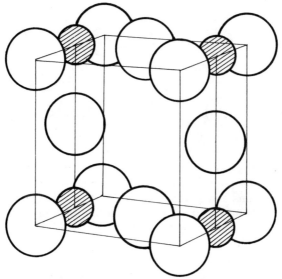

Fig. 22.   The aluminium fluoride lattice. 6:2 co-
ordination (Al^{3+} ions shaded).

It follows from the first law (p. 72) that:

$$\Delta H_f = \Delta H \text{ (atomisation)} + \tfrac{1}{2}\Delta H \text{ (dissociation)} + \text{(ionisation}$$
energy of metal) − (electron affinity of X) − U.

For any ionic solid the lattice energy, U, can be calculated
from the Born-Landé equation if the lattice geometry is known.
All the other terms in the equation can be measured experimentally except the electron affinity of the non-metal atom.
The Born-Haber cycle is therefore used principally for the
calculation of that quantity.

Example

Given the following thermochemical data, calculate the
electron affinity of the F atom, i.e. $-\Delta H°$ for:

$$F \text{ (g)} + e = F^- \text{ (g)}$$

		$\Delta H°/\text{kJ mol}^{-1}$
Rb (s)	= Rb (g)	+ 78
Rb (g)	= Rb^+ (g) + e	+402
$\tfrac{1}{2}F_2$ (g)	= F (g)	+ 80
$\tfrac{1}{2}F_2$ (g) + Rb (s)	= RbF (s)	−552
Rb^+ (g) + F^- (g)	= RbF (s)	−762

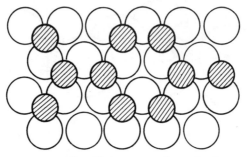

Fig. 23. The Al_2O_3 (corundum) lattice:
6:4 coordination. Adjacent layers. Al^{3+}
ions are shaded. In the corresponding
layers of the NaCl lattice the lines of
cations are unbroken.

One method of working such an example is to write down the
five given equations in such an order that, on addition, the
required equation is obtained:

		$\Delta H° / kJ\,mol^{-1}$	Notes
F (g)	$= \frac{1}{2}F_2(g)$	-80	F (g) required on left. Note change of sign on reversing order of equations.
$\frac{1}{2}F_2$ (g) + Rb (s)	= RbF (s)	-552	Eliminating unwanted $\frac{1}{2}F_2$ (g).
Rb (g)	= Rb (s)	-78	Eliminating unwanted Rb (s).
Rb^+ (g) + e	= Rb (g)	-402	Eliminating Rb (g).
RbF (s)	$= Rb^+$ (g) + F^- (g)	$+762$	Eliminating RbF (s) and Rb^+(g).

F (g) + e	$= F^-$ (g)	-350	

Thus the electron affinity for the F atom, the heat released
when it gains an electron, is +350 kJ mol^{-1}.

Examples V

1. Reflections of first, second and third order from one face of
a KCl crystal were obtained at glancing angles of 5.30°, 10.65°
and 16.10° respectively using X-rays of wavelength 58.5 pm.
Calculate the distance between successive planes of ions
parallel to the face.

2. From one face of a MgO crystal, X-rays of wavelength 58.5 pm gave first- and second-order reflections at glancing angles of 8° 12' and 16° 27' respectively. Calculate the distance between successive planes parallel to the face.

3. Calculate the force between a point charge $+3e$ and a point charge $-2e$ separated by 525 pm in a vacuum.

4. Calculate the energy released when a point charge $+3e$ and a point charge $-2e$ approach from infinite distance to 525 pm from one another.

5. Use the Born-Landé equation to calculate the lattice energy of calcium oxide, which has the NaCl lattice. Take n to be 8 and $r_0 = 239$ pm.

6. Calculate the lattice energy of zinc oxide (würtzite lattice) using the Born-Landé equation. (Zn^{2+} has the $3d^{10}$ structure, and $r_0 = 195$ pm).

7. Calculate the radius of the smallest sphere placed in the space A which would touch all three of the equal spheres in contact.

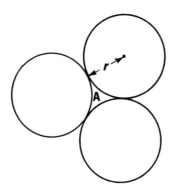

8. Calculate the radius of the sphere which would be just large enough to occupy the centre of a cube of length $2\,r$ and be just big enough to touch eight spheres in contact with one another with their centres at the corners of the cube. Hence calculate the smallest possible radius ratio which permits 8-co-ordination in an ionic crystal.

9. Spheres of radius $\sqrt{2}\,r$ at diagonally opposite corners of one face of a cube of side $2\,r$ both touch a sphere at the cube centre. Calculate its radius. Hence calculate the smallest radius ratio which permits 4-co-ordination in an ionic crystal.

10. Calculate the ionic radii for K^+ $(Z = 19)$ and Cl^- $(Z = 17)$, both with the $1s^2\ 2s^2\ 2p^6\ 3s^2\ 3p^6$ structure, for which the Slater screening constant $s = 11.6$, given that $r_{K^+} + r_{Cl^-}$, from X-ray measurements, is 314 pm.

11. Calculate the ionic radii for the isoelectronic ions Rb^+ and Br^- given that $(r_{Rb^+} + r_{Br^-}) = 348$ pm. ($s = 28.1$ in each case).

12. Use the answers obtained in 10 and 11 above to calculate the probable distances between adjacent nuclei in (a) KBr and (b) RbCl.

13. There is evidence that the iodide ions in TlI are just in contact with one another. TlI has the CsCl lattice (see 8 above) and $r_{I^-} = 216$ pm. What is the probable size of r_{Tl^+}?

14. The density of solid caesium chloride is 3.97×10^3 kg m^{-3}. $M_{CsCl} = 0.168$ kg mol^{-1} and the $Cs^+ - Cl^-$ distance in the crystal is 357 pm. A cube of side $\dfrac{2 \times 357}{3}$ pm therefore contains one pair of ions. Calculate N_A from the data.

15. Calculate N_A from the following data for CaF_2. $M_{CaF_2} = 0.0781$ kg mol^{-1}, density $= 3.18 \times 10^3$ kg m^{-3}, the $Ca^{2+} - F^-$ distance $= 237$ pm. (A cube of side $\dfrac{4 \times 237}{\sqrt{3}}$ pm contains four Ca^{2+} and eight F^- ions).

16. Calculate the lattice energy of AlF_3 (s) from:

		$\Delta H°$ / kJ mol^{-1}
Al (s) $+ \frac{3}{2}F_2$ (g)	$= AlF_3$ (s)	-1310
Al (s)	$= Al$ (g)	$+ 326$
F_2 (g)	$= 2\ F$ (g)	$+ 160$
F (g) $+$ e	$= F^-$ (g)	$- 350$
Al (g)	$= Al^{3+}$ (g) $+ 3e$	$+ 5138$

17. Calculate $\Delta H°$ for O (g) $+ 2e = O^{2-}$ (g) from:

		$\Delta H°$ / kJ mol^{-1}
Ba (s) $+ \frac{1}{2}O_2$ (g)	$= BaO$ (s)	$- 558$
Ba (s)	$= Ba$ (g)	$+ 178$
Ba (g)	$= Ba^{2+}$ (g) $+ 2e$	$+ 1465$
O_2 (g)	$= 2\ O$ (g)	$+ 498$
Ba^{2+} (g) $+ O^{2-}$ (g)	$= BaO$ (s)	-3198

18.   Calculate the electron affinity of the Cl atom, A_{Cl}, i.e. $-\Delta H°$ for $Cl + e = Cl^-$, from:

$$\Delta H°/kJ \ mol^{-1}$$

Rb (s) + $\frac{1}{2}Cl_2$ (g)	= RbCl (s)	−431
Rb (s)	= Rb (g)	+ 78
Rb (g)	= Rb^+ (g) + e	+402
Cl_2 (g)	= 2 Cl (g)	+242
Cl^- (g) + Rb^+ (g)	= RbCl (s)	−672

19.   For NaF (s), $U = 894$ kJ mol^{-1} and for NaCl (s), $U = 768$ kJ mol^{-1}. Use Table XIV (p. 76) and the electron affinities,

$$Cl \ (g) + e = Cl^- \ (g) \quad \Delta H° = -348 \ kJ \ mol^{-1}$$
$$F \ (g) + e = F^- \ (g) \quad \Delta H° = -352 \ kJ \ mol^{-1}$$

to calculate $\Delta H°$ for the reaction

$$2 \ NaCl \ (s) + F_2 \ (g) = 2 \ NaF \ (s) + Cl_2 \ (g)$$

20.   Use the data given in Q. 19 and the fact that the bond dissociation energy of the C−Cl bond in CH_3Cl is 338 kJ mol^{-1}, to calculate the bond dissociation energy of the C−F bond in CH_3F from

$$CH_3Cl + NaF = CH_3F + NaCl \quad \Delta H° = -4 \ kJ \ mol^{-1}.$$

Chapter 6

Thermochemistry

➤ Most chemical reactions are exothermic; they are accompanied by the production of heat. The reaction:

$$C \text{ (s)} + O_2 \text{ (g)} = CO_2 \text{ (g)}$$

for example, is that mainly responsible for heat production in a coke fire. Spontaneous endothermic reactions, in which heat is absorbed, are much less common. One example is:

$$Al_2(SO_4)_3 \text{ (aq)} + 3 \text{ Na}_2CO_3 \text{ (aq)} + 3 \text{ H}_2O \text{ (l)} = 2 \text{ Al(OH)}_3 \text{ (s)}$$
$$+ 3 \text{ Na}_2SO_4 \text{ (aq)} + 3 \text{ CO}_2 \text{ (g)}.$$

➤ The branch of chemistry which deals with heat changes in reactions is thermochemistry. Note that in thermochemical equations the physical state of each reactant and product must be specified.

Calorimetry

To measure the heat gained or lost in a reaction it is often sufficient to mix together in a calorimeter stoichiometric quantities of the reactants and to measure the resultant change in temperature.

Example

0.050 mol $Al_2(SO_4)_3$ in 0.050 kg of water and 0.150 mol Na_2CO_3 in 0.150 kg of water, both originally at 293.05 K, after thorough mixing in a calorimeter with a heat capacity of 155.0 J K^{-1}, cooled to 291.75 K.

Taking the heat capacity of the final solution to be that of the water, 4.184 kJ K^{-1} kg^{-1}, the total heat absorbed is given by:

70

q = (heat capacity of water + calorimeter) × (fall in temperature)

$$= [(0.200 \text{ kg} \times 4.184 \times 10^3 \text{ J K}^{-1} \text{ kg}^{-1}) + 155.0 \text{ J K}^{-1}]$$
$$\times (293.05 - 291.75) \text{ K} = 1290 \text{ J}.$$

For one mole of reaction as expressed in the equation, i.e. between one mole of $Al_2(SO_4)_3$ and three moles of Na_2CO_3, the heat absorbed is 1290 J/0.050 mol = 25.8 kJ mol^{-1}.

The thermochemistry of solutions is somewhat dependent on their concentrations. The addition of (aq) to a formula refers, strictly, to a very dilute solution (p. 77).

A particularly useful type of measurement in thermochemistry is that of the heat evolved when an element or compound burns in an excess of oxygen. The apparatus consists essentially of a small, steel pressure vessel, or 'bomb', which can be filled with oxygen at about 10^6 N m^{-2} pressure, and in which a sample can be ignited electrically.

Example

The combustion of a specimen of pure graphite weighing 0.600 g in a bomb calorimeter immersed in water raises the temperature of the calorimeter and water (of total heat capacity 9.42 kJ K^{-1}) by 2.09 K.

The heat evolved is, therefore:

$$9.42 \text{ kJ K}^{-1} \times 2.09 \text{ K} = 19.7 \text{ kJ}.$$

For one mole of reaction, therefore, the heat evolved is:

$$\frac{12.00 \text{ g mol}^{-1}}{0.600 \text{ g}} \times 19.7 \text{ kJ} = 394 \text{ kJ mol}^{-1}.$$

Another useful apparatus is a gas calorimeter in which gas moving at a measured rate through a tube burns at a jet and heats water moving at a measured rate through a coil surrounding the flame.

Enthalpy

Experiment shows that it is always possible to assign to any substance under specified physical conditions a quantity, H, the enthalpy, such that in any reaction at constant temperature and pressure, ΔH, i.e. $H_{products} - H_{reactants}$, is equal to the heat absorbed from the surroundings. By convention an element in its stable form at 298 K and 101 325 N m^{-2} – its so-called standard state – has zero enthalpy.

Thus $H°$ (α-sulphur) = 0, $H°$ (graphite) = 0, $H°$ O_2 (g) = 0.

If one mole of graphite and one mole O_2 in their standard states react to give carbon dioxide, 393 kJ of heat are evolved:

thus for C (s) + O_2 (g) = CO_2 (g) $\Delta H°$ = -393 kJ mol^{-1}.

The enthalpy of this reaction is also the standard enthalpy of formation of CO_2, $H°$ (CO_2), since it refers to its production from its elements for each of which $H°$ is zero.

In this book we shall avoid the use of terms like 'heat of reaction' and 'heat of formation'. Unfortunately some of these terms are traditionally applied to heat absorbed and others to heat evolved, causing considerable confusion. The term enthalpy of reaction always refers to $H_{\text{products}} - H_{\text{reactants}}$; i.e. a negative value of ΔH indicates an exothermic reaction.

A substance for which the standard enthalpy of formation $H°$ has a negative value is an exothermic compound; one for which $H°$ has a positive value is an endothermic compound.

The first law of thermodynamics

It is found by experiment that though one form of energy can be converted into another (e.g. work into heat, nuclear binding energy into kinetic energy), energy cannot be created or destroyed. A system such as a calorimeter and its contents has an enthalpy which depends on its temperature, pressure and composition. If some change, chemical or physical, occurs in the system, the enthalpy at the end of the process also depends entirely on the temperature, pressure and composition. Thus the enthalpy change of a process,

$$\Delta H = H_{\text{final}} - H_{\text{original}},$$

is independent of the way the process occurs.

For example, if ammonia is neutralised by nitric acid and then diluted with water to give very dilute NH_4NO_3, $\Delta H°$ = -119 kJ mol^{-1}:

		$\Delta H°/$ kJ mol^{-1}
NH_3 (g) + HNO_3 (l)	= NH_4NO_3 (s)	-145
NH_4NO_3 (s) + aq	= NH_4NO_3 (aq)	$+ 26$
NH_3 (g) + HNO_3 (l) + aq = NH_4NO_3 (aq)		-119

As expected from the first law, the enthalpy change is also $\Delta H°$ = -119 kJ mol^{-1} if the ammonia and nitric acid are diluted first and then mixed.

$\Delta H°/\text{kJ mol}^{-1}$

NH_3 (g) + aq	$= NH_3$ (aq)	-35
HNO_3 (l) + aq	$= HNO_3$ (aq)	-31
NH_3 (aq) + HNO_3 (aq)	$= NH_4NO_3$ (aq)	-53

NH_3 (g) + HNO_3 (l) + aq = NH_4NO_3 (aq)	-119

This aspect of the first law, that the enthalpy change of a chemical reaction is independent of the way the reaction is carried out, is known as Hess' law of heat summation (1840). It enables us to calculate the standard enthalpy of a compound like carbon monoxide which cannot be made directly from its elements in a calorimetry experiment. When the gas is converted to CO_2 in a gas calorimeter $\Delta H° = -284$ kJ mol^{-1}, and when graphite is converted to CO_2 in a bomb calorimeter $\Delta H° = -393$ kJ mol^{-1}.

Thus:

$\Delta H°/\text{kJ mol}^{-1}$

C (s) + O_2 (g)	$= CO_2$ (g)	-393
CO_2 (g)	$= CO$ (g) + $\frac{1}{2}O_2$ (g)	$+284$

C (s) + $\frac{1}{2}O_2$ (g) = CO (g)	-109

Note that if it is necessary to reverse an equation the sign of ΔH must be changed (Laplace's law). Similarly ΔH for n moles of reaction is $n \times \Delta H$ for one mole of reaction.

The standard enthalpy of a compound can often be calculated from its enthalpy of combustion, that is the increase in enthalpy which accompanies the complete combustion of unit amount of the compound in oxygen. Thus for

C_3H_8 (g) + 5 O_2 (g) = 3 CO_2 (g) + 4 H_2O (l) $\Delta H° = -2215$ kJ mol^{-1}.

Using this information, and the fact that $H°$ (H_2O, l) $= -285$ kJ mol^{-1} and $H°$ (CO_2, g) $= -393$ kJ mol^{-1}, the calculation can be performed as follows:

$\Delta H°/\text{kJ mol}^{-1}$

3 C (s) + 3 O_2 (g)	$= 3 CO_2$ (g)	-1179	$(3 \times H° (CO_2))$
4 H_2 (g) + 2 O_2 (g)	$= 4 H_2O$ (l)	-1140	$(4 \times H° (H_2O))$
3 CO_2 (g) + 4 H_2O (l)	$= C_3H_8$ (g) + 5 O_2 (g)	$+2215$	

3 C (s) + 4 H_2 (g)	$= C_3H_8$ (g)	-104

Thus $H°$ (C_3H_8, g) $= -104$ kJ mol^{-1}.

Constant-volume calorimetry

Most calorimetry experiments are performed at atmospheric pressure, and ΔH (reaction) is defined as the heat absorbed at constant pressure and temperature. But in a bomb calorimeter, a pressure change can occur during a combustion. In the reaction

$$C_{10}H_8 \text{ (s)} + 12 \ O_2 \text{ (g)} = 10 \ CO_2 \text{ (g)} + 4 \ H_2O \text{ (l)}$$

the amount of gas is reduced by 2 moles. Suppose the rise in temperature shows a quantity of heat q to be evolved, $\Delta H°$ is not simply $-q$ as in a reaction at constant pressure, but

$$\Delta H° = -q + V\Delta P$$
$$= -q + RT \times (-2 \text{ mol}),$$

if we assume the gases to behave ideally.

At a temperature of 298 K

$$\Delta H° = -q - 2 \text{ mol} \times 8.314 \text{ J K}^{-1} \text{ mol}^{-1} \times 298 \text{ K}$$
$$= -q - 4.9 \text{ kJ}$$

In general, for a reaction occurring at constant volume in which the amount of gas increases by Δn:

$$\Delta H° = -q + \Delta n RT$$

Bond dissociation energy

When I_2 gas is heated it dissociates:

$$I_2 \text{ (g)} \rightleftharpoons 2 \ I \text{ (g)}$$

and it is possible to calculate $\Delta H°$ for the forward reaction from measurements of the equilibrium constant at various temperatures (p. 96). As the energy absorbed in the reaction is all used in breaking the I-I bonds it is called the bond dissociation energy for the bond. Bond dissociation energies have been determined accurately for diatomic molecules by various methods outside the scope of this book.

Standard enthalpies of gaseous atoms

For the reaction:

$$I_2 \text{ (g)} = 2 \ I \text{ (g)} \quad \Delta H° = +150 \text{ kJ mol}^{-1},$$

which is the dissociation energy of the I-I bond.

Thus for $\frac{1}{2} I_2 \text{ (g)} = I \text{ (g)} \quad \Delta H° = +75 \text{ kJ mol}^{-1}.$

But in its standard state iodine is a solid, and for

$$\tfrac{1}{2}I_2 \text{ (s)} = \tfrac{1}{2}I_2 \text{ (g)} \quad \Delta H° = +31 \text{ kJ mol}^{-1}.$$

Thus, as $H°$ (I_2, s) = 0, the element being in its standard state,

$$H° \text{ (I, g)} = (+75 + 31) \text{ kJ mol}^{-1}$$
$$= +106 \text{ kJ mol}^{-1}.$$

Table XIV lists the standard enthalpies of formation of some atoms from the elements in their standard states.

Mean bond energies

For H_2O gas, $H° = -241 \text{ kJ mol}^{-1}$.

Using Table XIV we can calculate the enthalpy of formation of the gas molecule from its atoms:

	$\Delta H°/\text{kJ mol}^{-1}$
H_2 (g) $+\tfrac{1}{2}O_2$ (g) $= H_2O$ (g)	-241
2 H (g) $\quad\quad = H_2$ (g)	-435
O (g) $\quad\quad = \tfrac{1}{2}O_2$ (g)	-248
2 H (g) + O (g) $= H_2O$ (g)	-924

Representing the reaction of gaseous atoms to give a gaseous molecule as

$$H + O + H = H - O - H$$

we see that the release of energy arises solely from the formation of two O-H bonds. The mean energy release in the formation of an O-H bond, the mean bond energy, E (O-H), is therefore $924/2 \text{ kJ mol}^{-1} = 462 \text{ kJ mol}^{-1}$.

Experiments show that the strength of a particular bond varies little from one compound to another. For example, the O-H bonds in H_2O_2 can be expected to have much the same strength as those in H_2O. This assumption enables us to calculate E (O-O) from $H°$ (H_2O_2, g):

	$\Delta H°/\text{kJ mol}^{-1}$
H_2 (g) + O_2 (g) $\quad = H_2O_2$ (g)	$- 130$
2 H (g) $\quad\quad = H_2$ (g)	$- 435$
2 O (g) $\quad\quad = O_2$ (g)	$- 496$
2 H (g) + 2 O (g) $= H_2O_2$ (g)	-1061

Thus for $H + O + O + H = H - O - O - H$ the energy release is 1061 kJ mol^{-1}. But the formation of two O-H bonds releases

TABLE XIV
Standard Enthalpies of Some Gaseous Atoms

	$\Delta H° \,/\, \text{kJ mol}^{-1}$
C (s)　　 = C (g)	+ 715
$\frac{1}{2}H_2$ (g)　 = H (g)	+ 217.5
$\frac{1}{2}O_2$ (g)　 = O (g)	+ 248
$\frac{1}{2}F_2$ (g)　 = F (g)	+ 80
$\frac{1}{2}Cl_2$ (g) = Cl (g)	+ 121
$\frac{1}{2}Br_2$ (l) = Br (g)	+ 112
$\frac{1}{2}I_2$ (s)　 = I (g)	+ 106
$\frac{1}{2}N_2$ (g) = N (g)	+ 472.5

$2 \times E$ (O–H) $= 924$ kJ mol^{-1}. Thus E (O–O) $= (1061 - 924)$ kJ mol^{-1} $= 137$ kJ mol^{-1}.

Resonance energy

The structural formulae of many compounds, particularly carbon compounds, can be represented fairly realistically using single, double and triple bonds. Thus it is possible to calculate values of E (C = C) and E (C ≡ C), for example, which are consistent with the enthalpies of gaseous alkenes and alkynes. But where the structural formula is not a realistic representation of the bonding in the molecule, as in the Kekulé formula of benzene:

the true standard enthalpy of formation is less than the value calculated from the simple formula, i.e. the molecule is more thermodynamically stable. The difference (calculated standard enthalpy – experimental standard enthalpy) is called, for historical reasons which need not concern us here, the reso- nance stabilisation energy of the molecule.

Enthalpy of solution and enthalpy of dilution

The integral enthalpy of solution for one mole of a compound is the heat absorbed at constant temperature and pressure when the compound is dissolved in a very large volume of water; the heat absorbed depends on the concentration of the solution produced, but becomes a constant for this case.

The reaction is written for the compound MX:

$$MX + aq = MX \text{ (aq)}$$

where MX (aq) represents the very dilute solution.

The enthalpy of dilution is the enthalpy change when water is added to a solution. The concentration of both solutions must be specified; for example:

$$MX \ (conc. \ 10^3 \ mol \ m^{-3}) + aq = MX \ (conc. \ 10 \ mol \ m^{-3}).$$

Enthalpies of solution are often easy to measure and can be used in the calculation of standard enthalpies of compounds. An example is given below:

$$\Delta H°/kJ \ mol^{-1}$$

NaOH (aq)	= NaOH (s) + aq	+ 43
Na (s) + H_2O (l) + aq	= NaOH (aq) + $\frac{1}{2}H_2$ (g)	−184
H_2 (g) + $\frac{1}{2}O_2$ (g)	= H_2O (l)	−285

$$Na \text{ (s)} + \tfrac{1}{2}H_2 \text{ (g)} + \tfrac{1}{2}O_2 \text{ (g)} = NaOH \text{ (s)} \qquad −426$$

Thus $H°$ (NaOH, s) = −426 kJ mol^{-1}.

Enthalpy of neutralisation

The neutralisation of aqueous acid by aqueous alkali is always exothermal, e.g:

$$HNO_3 \text{ (aq)} + NaOH \text{ (aq)} = NaNO_3 \text{ (aq)} + H_2O$$
$$\Delta H° = −57.3 \ kJ \ mol^{-1}.$$

The enthalpy change which accompanies the conversion of acid and base to salt plus one mole of H_2O is called the enthalpy of neutralisation. The value is close to constant for the neutralisation of any strong acid (p. 136) by any strong base. In the example just given, the reaction can be represented more realistically by the equation

$$H^+ \text{ (aq)} + NO_3^- \text{ (aq)} + Na^+ \text{ (aq)} + OH^- \text{ (aq)}$$
$$= H_2O \text{ (l)} + Na^+ \text{ (aq)} + NO_3^- \text{ (aq)}$$

i.e., since the Na^+ (aq) and NO_3^- (aq) ions remain unchanged,

$$H^+ \text{ (aq)} + OH^- \text{ (aq)} = H_2O + \text{aq}.$$

This last equation represents the true reaction in the neutral-isation of any fully ionised acid by a fully ionised base, and the enthalpy change has always the same value. But for the neutral-isation of, say, a weak acid by a strong base the reaction is:

$$HA \text{ (aq)} + OH^- \text{ (aq)} = H_2O + A^- \text{ (aq)}$$

and the enthalpy change depends on the nature of HA.

Such reactions are less exothermic than those between strong acids and strong bases because energy is needed to dissociate a weak acid or base into ions.

Examples VI

1. The combustion of 1.22 g of benzoic acid in a bomb calori-meter raised the temperature of the water surrounding it by 3.86 K. The heat capacity of water and calorimeter was 8364 $J K^{-1}$. Calculate the heat evolved per mole of reaction:

$$C_6H_5CO_2H \text{ (s)} + 7\tfrac{1}{2}O_2 \text{ (g)} = 7\ CO_2 \text{ (g)} + 3\ H_2O \text{ (l)}$$

2. When 100.0 cm^3 KOH containing 1.00 kmol m^{-3} and 100.0 cm^3 HCl of the same concentration were mixed in a calorimeter of heat capacity 92 $J K^{-1}$ the temperature was raised by 6.17 K. Calculate the heat evolved per mole of reaction:

$$KOH \text{ (aq)} + HCl \text{ (aq)} = KCl \text{ (aq)} + H_2O \text{ (l)}$$

3. When 100.0 cm^3 KOH containing 1.00 kmol m^{-3} and 100.0 cm^3 CH_3CO_2H of the same concentration were mixed in the same calorimeter above, the temperature rise was 5.18 K. Calculate the heat evolved per mole of reaction:

$$KOH \text{ (aq)} + CH_3CO_2H \text{ (aq)} = CH_3CO_2K \text{ (aq)} + H_2O \text{ (l)}$$

4. In a gas calorimeter, butane passing through a tube at 1.00×10^{-4} mol s^{-1} is burnt at a jet. The temperature of water passing through a coil surrounding the jet at the rate of 20.0 g s^{-1} is raised 3.44 K by the flame. Calculate the heat evolved in the combustion of one mole of butane.

5. Given that $H°$ (H_2O liquid) $= -285$ kJ mol^{-1} and that $H°$ (CO_2 gas) $= -393$ kJ mol^{-1}, calculate

    (a) $H°$ (n-C_5H_{12} gas) from

$$C_5H_{12} \text{ (g)} + 8\ O_2 \text{ (g)} = 5\ CO_2 \text{ (g)} + 6\ H_2O \text{ (l)}$$
$$\Delta H° = -3529 \text{ kJ mol}^{-1}.$$

(b) $H°$ ($H_2C_2O_4$) from:

$$H_2C_2O_4 \text{ (s)} + \tfrac{1}{2}O_2 \text{ (g)} = 2\ CO_2 \text{ (g)} + H_2O \text{ (l)}$$
$$\Delta H° = -146 \text{ kJ mol}^{-1}.$$

(c) $H°$ ($(NH_2)_2CO$) from:

$$(NH_2)_2CO \text{ (s)} + \tfrac{3}{2}O_2 \text{ (g)} = N_2 \text{ (g)} + CO_2 \text{ (g)} + 2\ H_2O \text{ (l)}$$
$$\Delta H° = -630 \text{ kJ mol}^{-1}.$$

6. $H°$ (n-C_4H_{10}, g) $= -124$ kJ mol^{-1} and $H°$ (CH_4, g) $= -63$ kJ mol^{-1}. Calculate $\Delta H°$ for the reaction:

$$C_4H_{10} \text{ (g)} + 3\ H_2 \text{ (g)} = 4\ CH_4 \text{ (g)}$$

7. $H°$ (C_2H_2, g) $= +226$ kJ mol^{-1}. $H°$ (C_2H_6, g) $= -84$ kJ mol^{-1}. Calculate $\Delta H°$ for the reaction:

$$C_2H_2 \text{ (g)} + 2\ H_2 \text{ (g)} = C_2H_6 \text{ (g)}$$

8. Calculate ΔnRT for the combustion of succinic acid:

$$C_4H_6O_4 \text{ (s)} + 3\tfrac{1}{2}O_2 \text{ (g)} = 4\ CO_2 \text{ (g)} + 3\ H_2O \text{ (l)}$$

at 298 K and hence find the quantity by which $\Delta H°$ for the reaction exceeds the heat absorbed during the combustion in a bomb calorimeter.

9. For CH_4 gas, $H° = -63$ kJ mol^{-1}. Use this information and the values for $H°$ (H) and $H°$ (C gas) in Table XIV to calculate the mean bond energy of the C–H bond.

10. For C_2H_6 gas, $H° = -84$ kJ mol^{-1}. Use this information, Table XIV, and the answer to Q. 9 above to calculate E (C–C).

11. For acetylene gas, C_2H_2, $H° = +226$ kJ mol^{-1}. Calculate E (C $\equiv$ C).

12. For NH_3 gas, $H° = -45$ kJ mol^{-1}. Use this information and Table XIV to calculate (a) E (N–H). Hence find (b) E (N–N) in N_2H_4 given that $H°$ (N_2H_4 gas) $= +50$ kJ mol^{-1}.

13. For CH_2Cl_2 gas, $H° = -88$ kJ mol^{-1}. From Table XIV and the value for E (C–H) obtained in Q. 9, find the value of E (C–Cl) in this compound.

14. For CH_2O gas, $H° = -116$ kJ mol^{-1}. Calculate the value of E (C = O).

15. For the reaction C (s) + 2 Br_2 (l) = CBr_4 (g), $\Delta H° = +50$ kJ mol^{-1}. Use this information, and Table XIV, to calculate E (C–Br).

16. For gaseous benzene, $H°$ (C_6H_6) = +83 kJ mol^{-1}. Given that E (C = C) = +615 kJ mol^{-1}, and using Table XIV and the values of E (C-H) and E (C-C) calculated above, find the resonance stabilisation energy of benzene.

17. Calculate the value of $H°$ expected for formic acid gas, assuming its formula to be H-C$\diagup$OH$\diagdown$O , taking E (C-O) to be +360 kJ mol^{-1}, using Table XIV and the values of E (C-H), E (C=O) and E (O-H) (in text) calculated earlier. The experimental value of $H°$ (formic acid gas) = -363 kJ mol^{-1}. Hence calculate the resonance stabilisation energy of the molecule.

18. Calculate $H°$ (KOH, s) from the following data:

$\Delta H°/\text{kJ mol}^{-1}$

KOH (s) + aq = KOH (aq)	$-$ 55
H_2 (g) + $\frac{1}{2}O_2$ (g) = H_2O (l)	-285
K (s) + H_2O (l) + aq = KOH (aq) + $\frac{1}{2}H_2$ (g)	-195

19. Calculate $\Delta H°$ for the reaction

$$\text{Ca (s)} + \tfrac{1}{2}O_2 \text{ (g)} = \text{CaO (s)}$$ from the data:

$\Delta H°/\text{kJ mol}^{-1}$

CaO (s) + H_2O (l) + aq = Ca(OH)$_2$ (aq)	$-$ 81
Ca (s) + 2 H_2O (l) + aq = Ca(OH)$_2$ (aq) + H_2 (g)	-431
H_2 (g) + $\frac{1}{2}O_2$ (g) = H_2O (l)	-285

20. Calculate $H°$ (HI, g) from the data:

$\Delta H°/\text{kJ mol}^{-1}$

HI (g) + aq = HI (aq)	$-$ 80
HI (aq) + KOH (aq) = KI (aq)	$-$ 57
KI (aq) + $\frac{1}{2}Cl_2$ (g) = KCl (aq) + $\frac{1}{2}I_2$ (s)	-109
HCl (aq) + KOH (aq) = KCl (aq)	$-$ 58
$\frac{1}{2}H_2$ (g) + $\frac{1}{2}Cl_2$ (g) = HCl (g)	$-$ 92
HCl (g) + aq = HCl (aq)	$-$ 72

Chapter 7

Chemical equilibria

Some chemical reactions proceed to completion, e.g:

$$Zn + 2 H^+ \text{ (aq) in excess} = Zn^{2+} + H_2,$$

but others, e.g:

$$CH_3COOH + C_2H_5OH = CH_3COOC_2H_5 + H_2O$$

appear to stop at a stage where appreciable amounts of the reactants remain unchanged. A state of chemical equilibrium is reached in this case as the result of the balancing of the two opposing reactions. When equilibrium is reached for this reaction at a particular temperature the relation between the activities (p. 89) of reactants and products is:

$$\frac{a_{CH_3COOC_2H_5} \times a_{H_2O}}{a_{CH_3COOH} \times a_{C_2H_5OH}} = K,$$

where K is the equilibrium constant for the reaction, as written above, at that temperature. Exactly the same relation between the respective activities is obtained whether we start by mixing acetic acid and ethanol or by mixing ethyl acetate and water.

Although K is often so large that a reaction appears to go to completion, every reaction in which the products remain in contact with the reactants does eventually reach an equilibrium state. However, if one of the products is removed from the system, as, for example, when it escapes as a gas:

$$Cl^- \text{ (aq)} + H_2SO_4 = HCl \text{ (g)} + HSO_4^- \text{ (aq)},$$

the reaction continues until one of the reactants is completely consumed.

The equilibrium equation

For the general equilibrium reaction

$$aA + bB \rightleftharpoons cC + dD$$

the equilibrium constant is given by

$$K = \frac{(a_C)^c \times (a_D)^d}{(a_A)^a \times (a_B)^b} \qquad (36)$$

Note that we replace the equals sign by two arrows in the equation of reaction and also that, in expressing K, we place activities of <u>products</u> (on the right of the equation of reaction) in the numerator and activities of <u>reactants</u> in the denominator.

To explain the significance of Eqn. (36) Guldberg and Waage (1864) argued that the rate of the forward reaction was $k_f (a_A)^a (a_B)^b$ and that of the back reaction $k_b (a_C)^c (a_D)^d$. As the concentrations, and therefore the activities, of C and D are built up, the speed of the back reaction increases, and eventually the forward and back reactions proceed at the same rate, i.e:

$$k_f \times (a_A)^a \times (a_B)^b = k_b \times (a_C)^c \times (a_D)^d$$

and

$$\frac{k_f}{k_b} = \frac{(a_C)^c \times (a_D)^d}{(a_A)^a \times (a_B)^b} = K.$$

Unfortunately this attractively simple explanation is based on the assumption that the rate of a reaction:

$$aA + bB \longrightarrow products$$

is necessarily proportional to $(a_A)^a \times (a_B)^b$, and though this is sometimes true, it is not always so by any means (p. 110). A more satisfactory explanation of Eqn. (36) arises from the <u>second law of thermodynamics</u>. This is a law of experience, which means it is based on experiment, and it is so fundamental that we have all accepted its implications from early childhood.

Before we deal with the formal statement of the law, it is necessary to introduce the concept of <u>entropy</u>, and we shall attempt to give physical significance to this quantity by means of analogies.

The earliest workers in the field of thermochemistry (p. 70) thought that the reactive power of a substance was

determined by its enthalpy (p. 71), and that the driving force
of a reaction was its ability to produce heat. The explanation
was unsatisfactory because it failed to account for (a) spontaneous endothermic reactions and (b) chemical equilibria. We
now realise that the enthalpy change of a chemical process
provides only part of the driving force and that the other factor
of importance is the change of entropy dealt with in succeeding
paragraphs.

Entropy as the capacity factor of heat energy

If we apply a difference of electric potential U between
two ends of a metallic conductor, and a quantity of electric
charge, Q, flows in it, the electrical work done in the conductor
is

$$W = U \times Q \qquad (37)$$

In SI units, joules = volts × coulombs. Equation (37) is an
example of the general principle that energy is the product of
an intensity factor — some kind of 'push' — here the potential
difference, and a capacity factor — an 'amount pushed' — here
the amount of electric charge. Physically, of course, the flow
of what we called electricity consists really of the movement of
electrons through the metal.

It is instructive for our purposes to compare the transfer
of 'electricity' with the transfer of 'heat'. Let us imagine a
source at a temperature T (> 273 K) placed in thermal contact
with an ice calorimeter (at 273 K) through a thermal conductor
(Fig. 24). The difference in temperature, $T - 273$ K, provides
the 'push' for the process of thermal conduction; it is the
intensity factor of heat energy. The heat transferred to the
calorimeter can be calculated from the amount of ice which is
melted in the process; its units in the SI are joules and it is
analogous to the work done in the conductor in the transference
of electricity.

We must ask ourselves what, in the conduction of heat,
corresponds to the movement of electrons in the conduction of
electricity. The particles in a metal at high temperature are
in a state of greater vibrational agitation, and are therefore
more disordered at a particular instant, than those in the same
metal at low temperature. Across the conductor in Fig. 24 we
can therefore picture a kind of gradient of disorder, and the
physical picture is of heat being transmitted by the communication of vibrational disorder from one particle to another through

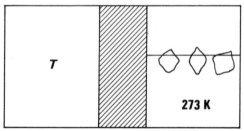

Fig. 24. Thermal conduction.

the metal. Furthermore, the disorder is eventually communi-
cated to the contents of the ice calorimeter, the ordered crystal
structure of ice being converted to the rather disordered
structure of liquid water as the result of the conduction process.
Thus the capacity factor of heat energy, physically analogous to
the electrons pushed through a conductor by a difference of
electric potential, is a kind of disorder pushed through the
thermal conductor by a difference of temperature. Furthermore since, analogous with Eqn. (37):

heat energy = temperature difference × the capacity factor
of heat energy,

it follows that the capacity factor — the 'disorder' factor — has
units in the SI of J K^{-1}.

Suppose that, in the experiment just described, an amount
of heat q is supplied to the ice calorimeter under reversible
conditions at a temperature T_1, then we define a property S of
the calorimeter by the equation

$$\Delta S = \frac{q_{rev}}{T} \tag{38}$$

Reversible conditions imply that only a very slight change in the
conditions would cause a reversal of the process. If we lower
the temperature around an ice–water calorimeter a very small
fraction of a kelvin below the ice-point, water is converted into
ice; if we raise it marginally above the ice-point, ice is converted to water. An example of a reversible process in a
chemical cell is given on p. 147.

We call the quantity ΔS in Eqn. (38) the increase in entropy
of the calorimeter. A convenient physical picture is that it
represents the increase in disorder which results from the
melting of the ice.

Entropy changes in fusion and vaporisation

The fusion of a solid at its normal melting point is a process which can be carried out, by its very nature, under reversible conditions at a constant temperature. If we supply 6.02 kJ of heat to a mixture of ice and water at 273 K, one mole of ice is melted; thus the entropy of the ice–water mixture is increased by:

$$\Delta S_f = \frac{q_{rev}}{T} = \frac{6.02 \times 10^3 \text{ J}}{273 \text{ K}}$$

$$= 22.0 \text{ J K}^{-1}.$$

Thus the entropy of fusion of ice at its melting point is 22.0 J K^{-1} mol^{-1}.

The vaporisation of a substance at its boiling point is also reversible. If 44.2 kJ of heat is supplied to a mixture of toluene and its vapour at 384 K, its boiling point, one mole of liquid is converted to vapour. Thus the increase in entropy of the liquid–vapour mixture is given by:

$$\Delta S_v = \frac{q}{T} = \frac{44.2 \times 10^3 \text{ J}}{384 \text{ K}}$$

$$= 115 \text{ J K}^{-1}$$

Thus the entropy of vaporisation of toluene is 115 J K^{-1} mol^{-1}. It is interesting that the molar entropy of vaporisation is commonly about 100 J K^{-1} mol^{-1} for a large range of substances which can be considered to contain simple molecules in both the liquid and vapour states. For liquids like water, ethanol and hydrogen fluoride, which are associated in the liquid phase because of hydrogen bonding, ΔS_v is appreciably larger; evidently the fact that the liquids have more ordered structures demands that more heat is required at the boiling point to increase the disorder of the state to that of the gaseous monomer.

Standard entropies

Just as every substance can be assigned a standard enthalpy, $H°$, (p. 72) based on the convention that $H°$ is zero for any element in its standard state, so every substance can be assigned a standard entropy, $S°$, based on the principle, known as the third law of thermodynamics, that a pure crystalline substance has zero entropy at zero on the thermodynamic

temperature scale. Various methods are available for the
experimental determination of standard entropies. For a substance which is gaseous at 298 K, the value of $S°$, usually given
for that temperature in tables, is much higher than that of a
liquid, which, in turn, is higher than that of a solid. For
example, at 298 K

$$S° \text{(graphite)} = 5.69 \text{ J K}^{-1} \text{ mol}^{-1}, \quad S° \text{(Br}_2, \text{l)} = 152 \text{ J K}^{-1} \text{ mol}^{-1},$$
$$S° \text{(O}_2, \text{g)} = 205 \text{ J K}^{-1} \text{ mol}^{-1}$$

These results are hardly surprising in view of what we have
said about the connection between entropy and disorder on p. 84.

For our purposes the most important use of $S°$ values is
that they enable the entropy change $\Delta S°$ of a chemical reaction to
to be calculated in an exactly similar manner to the calculation
of $\Delta H°$ from $H°$ values (p. 73).

Example

Calculate $\Delta S°$ for the reaction

$$2 \text{ H}_2 \text{ (g)} + \text{O}_2 \text{ (g)} = 2 \text{ H}_2\text{O (l)}$$

at 298 K and $1.013 \times 10^5 \text{ N m}^{-2}$. The values of $S°$ at this temper-
ature and pressure are: $S° \text{ (H}_2\text{O, l)} = 69.5 \text{ J K}^{-1} \text{ mol}^{-1}$, $S° \text{ (H}_2, \text{g)}$
$= 130.5 \text{ J K}^{-1} \text{ mol}^{-1}$, $S° \text{ (O}_2, \text{g)} = 205.0 \text{ J K}^{-1} \text{ mol}^{-1}$.

$$\begin{aligned}
\Delta S° &= 2 S° \text{ (H}_2\text{O, l)} - S° \text{ (O}_2, \text{g)} - 2 S° \text{ (H}_2, \text{g)} \\
&= (139.0 - 205.0 - 261.0) \text{ J K}^{-1} \text{ mol}^{-1} \\
&= -327.0 \text{ J K}^{-1} \text{ mol}^{-1}
\end{aligned}$$

The large negative value of $\Delta S°$ indicates that the reaction is
accompanied by a decrease in disorder (i.e. an increase in
order) which is to be expected since a liquid is formed from
gases and also because only two molecules of product are
formed from every three molecules of reactants.

The second law of thermodynamics

The law states that in any spontaneous change in an
isolated system there is an increase in entropy. This may not
appear particularly meaningful at first glance, but when we consider the relation between entropy and disorder we begin to
recognise it as a law of everyday experience. If we are shown
a film in which broken glass and a pool of liquid on the floor of
a room assemble themselves into a jug of milk and jump up on

SWAN
MYNYDDBACH
COMPREHENSIVE
SCHOOL FOR GIRLS
EDUCATION 87
COMMITTEE

to a table we know at once we are seeing an actual event being shown in reverse order of time. We just cannot accept that order arises spontaneously out of disorder. Pursuing the point about early childhood made on p. 82, the toddler building his bricks into towers and finding that the natural tendency of his towers is to fall down, is in effect teaching himself the second law.

For our immediate purposes the law has two important consequences.

(a) Consider a system, thermally insulated from its surroundings, in which a compartment A at a temperature T_2 is separated from a compartment B at T_1 by a thermally-conducting partition (Fig. 25). Suppose that as a result of thermal conduction through the partition the heat content of B increases by q and that of A decreases by q. Then the entropy change in the isolated system is $\frac{q}{T_1} - \frac{q}{T_2}$. Since this must be positive for a spontaneous change, $T_2 > T_1$. Thus the direction of spontaneous thermal conduction is from the hotter body to the colder one; this is the best-known physical aspect of the second law.

(b) Consider an insulated system in which a compartment A is separated from its surroundings B by thermally conducting walls (Fig. 26).

Suppose that as a result of a spontaneous chemical reaction in A the entropy of A increases by an amount ΔS and the heat content of B increases by q. Then, from the second law,

$$\Delta S + \frac{q}{T} > 0$$

For a chemical reaction at constant pressure,

$$q = -\Delta H \text{ (p. 71)}$$

Therefore the condition for spontaneous change in a system at constant pressure at a temperature T is that

$$\Delta S - \frac{\Delta H}{T} > 0,$$

i.e. that $\Delta H - T\Delta S$ is negative.

For a reversible change at constant temperature and pressure, that is when the system is in a state of equilibrium:

$$\Delta H - T\Delta S = 0.$$

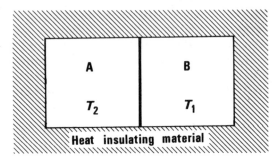

Fig. 25.   Direction of thermal conduction.

The Gibbs function *G*

We have seen above that the criterion for a spontaneous process at constant temperature and pressure is not that the enthalpy change, ΔH, should be negative, but that $\Delta H - T\Delta S$ should be negative. Thus a spontaneous endothermic process is possible provided ΔS for the process is positive and T is sufficiently large. For example, in the reaction,

$$N_2O_4 \text{ (g)} \rightleftharpoons 2\,NO_2 \text{ (g)},$$

ΔH for the forward reaction is positive because a nitrogen-nitrogen bond is broken (p. 74), and ΔS is positive for the forward reaction because there is an increase in disorder when one molecule of N_2O_4 splits into two molecules of NO_2. Thus a high temperature favours the forward reaction but a low temperature favours the back reaction.

It is obviously useful to define a property G for a system such that

$$G = H - TS \tag{39}$$

G is called the Gibbs function or Gibbs free energy of the system. The characteristic property of G is that, under conditions of constant temperature and pressure, a process can occur only in the direction of decreasing G. It follows that, under such conditions, an equilibrium state corresponds to a minimum in G, and the Gibbs function in chemical thermodynamics is analogous in this sense to potential energy in mechanics. Because of its importance, a very great deal of work has gone into compiling tables of standard values of G for chemical substances (cf. enthalpy tables, p. 72). There are several experimental methods available, and which is the most suitable depends on the compound. Values are usually quoted for the pure substance at 298 K and 101 325 N m^{-2}.

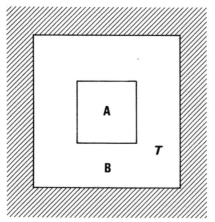

Fig. 26. Representation of system
and its surroundings.

In dealing with reactions we are concerned with the values
of G for mixtures rather than for pure substances. We use the
term chemical potential to signify the contribution made by any
constituent to the total Gibbs function of the mixture. In general
if the contribution made by one mole of a substance A is G_A,
then we define a quantity a_A, the relative activity of the substance A in that mixture at a given temperature and pressure,
by the equation

$$G_A = G°_A + RT \ln a_A \qquad (40)$$

where $G°_A$ is the Gibbs function for one mole of the pure substance A at the same temperature and pressure.

The relative activity is a dimensionless quantity, as the
form of the equation implies. A full treatment of the subject of
activity is outside the scope of this book, but a useful approximation for a very dilute solution of A is that

$$a_A = \frac{m_A}{\text{mol kg}^{-1}} \qquad (41)$$

The application of the concept to mixtures of ideal gases is
explained in the next paragraph, and to dilute solutions of ionic
substances on p. 131.

Equilibria in gaseous mixtures

In a mixture of gases A, B, C, each will contribute to the
total value of G. If we regard G_A as the contribution of unit

amount of A to the total G, its value is given to a close approximation, for conditions where the gas behaves approximately ideally, by

$$G_A = G°_A + RT \ln y_A \tag{42}$$

where $G°_A$ is the value of G_A for the pure gas at the same temperature and pressure, and y_A is the <u>mole fraction</u> of A in the mixture, i.e.

$$y_A = \frac{n_A}{n_A + n_B + n_C} \tag{43}$$

where n_A, n_B and n_C are the amounts of A, B and C as defined on p. 6, and where the symbols A, B and C refer to the molecular species which exist in the gas phase.

Let us now consider the general gas reaction

$$aA + bB \longrightarrow cC + dD.$$

Imagine one mole of (forward) reaction to occur in a total amount of gas so large that y_A, y_B, y_C and y_D are effectively unchanged in the process. Then the change in G in the process is given by

$$\begin{aligned} \Delta G &= c\,(G°_C + RT \ln y_C) + d\,(G°_D + RT \ln y_D) \\ &\quad - a\,(G°_A + RT \ln y_A) - b\,(G°_B + RT \ln y_B) \\ &= \Delta G° + RT \ln \frac{(y_C)^c\,(y_D)^d}{(y_A)^a\,(y_B)^b} \end{aligned} \tag{44}$$

Here $\Delta G°$ is simply $G°$ (products) $-G°$ (reactants) for one mole of reaction, and can be calculated from tables of standard Gibbs functions.

Consider now the case in which the values of y_A, y_B, y_C and y_D are the equilibrium values. If one mole of reaction occurs without sensibly altering the mole fractions, $\Delta G = 0$, as the system is effectively in equilibrium throughout.

Then from Eqn. (44), known as the <u>reaction isotherm</u>,

$$\Delta G° = -RT \ln K \tag{45}$$

where K is the equilibrium constant, expressed in terms of mole fractions, for the reaction at the temperature and pressure for which the $G°$ values apply.

Example

At 298 K and 1.013×10^5 N m^{-2}

$$G° \; (NO_2, \; g) \;\; = \; -37.8 \text{ kJ mol}^{-1},$$

$$G° \; (N_2O_4, \; g) \;\; = \; -81.3 \text{ kJ mol}^{-1}.$$

Calculate K for the reaction

$$N_2O_4 \; \rightleftharpoons \; 2 \; NO_2$$

under the same conditions.

$$
\begin{aligned}
\Delta G° \;\; &= \;\; 2 \; G° \; (NO_2) \; - \; G° \; (N_2O_4) \\
&= \;\; 2 \; (-37.8 \text{ kJ mol}^{-1}) \; - \; (-81.3 \text{ kJ mol}^{-1}) \\
&= \;\; +5.7 \text{ kJ mol}^{-1}.
\end{aligned}
$$

$$
\therefore 2.303 \; \log_{10} K \;\; = \;\; \frac{-5.7 \times 10^3 \text{ J mol}^{-1}}{8.31 \text{ J K}^{-1} \text{ mol}^{-1} \times 298 \text{ K}}
$$

$$\log_{10} K \;\; = \;\; -1.00 = \bar{1}.00$$

$$\therefore \; K \;\; = \;\; 0.100.$$

Dissociation in the gas phase

The above reaction is an example of gaseous dissociation. Suppose a fraction x of the N_2O_4 is dissociated into NO_2 at equilibrium, then

$$\frac{\text{amount of } NO_2}{\text{amount of } N_2O_4} \;\; = \;\; \frac{2x}{1-x} \;\; \text{at equilibrium.}$$

Thus

$$y_{NO_2} \;\; = \;\; \frac{2x}{(1-x)+2x} \;\; = \;\; \frac{2x}{1+x}$$

and

$$y_{N_2O_4} \;\; = \;\; \frac{1-x}{1+x} \;\; \text{at equilibrium.}$$

Therefore

$$K \;\; = \;\; \frac{\left(\dfrac{2x}{1+x}\right)^2}{\left(\dfrac{1-x}{1+x}\right)} \;\; = \;\; \frac{4x^2}{1-x^2}$$

Thus at 298 K and 101 325 N m^{-2}, the value of x is given by:

$$\frac{4x^2}{1-x^2} \;\; = \;\; 0.100$$

$$\therefore x^2 = \frac{0.100}{4.10}$$

$$\therefore x = 0.158.$$

The effect of pressure on gaseous equilibria

We have derived above an equation for the relation between the standard $G°$ values for the reactants and products and the equilibrium constant for the reaction under the condition of temperature and pressure for which the $G°$ values apply. If we replace Eqn. (42) by

$$G_A = G°_A + RT \ln \frac{p_A}{p°} \qquad (46)$$

however, in which p_A is the partial pressure (p. 44) of A in the mixture and $p°$ is the pressure at which the $G°$ value was determined, we obtain, by a derivation analogous to that on p. 90, an equation

$$\Delta G° = -RT \ln \frac{(p_C/p°)^c \times (p_D/p°)^d}{(p_A/p°)^a \times (p_B/p°)^b} \qquad (47)$$

for equilibrium conditions. We denote the fraction on the right by $K_{p/p°}$, the equilibrium constant in terms of partial pressures expressed as fractions of the standard pressure at which $G°$ is determined. $K_{p/p°}$, as so defined, is a dimensionless quantity.

Returning to the example

$$N_2O_4 \rightleftharpoons 2\ NO_2$$

we can see from Dalton's law (p. 44) and from p. 91 that if a fraction x of the N_2O_4 is dissociated

$$p_{N_2O_4} = \frac{1-x}{1+x} p,$$

and

$$p_{NO_2} = \frac{2x}{1+x} p,$$

where p is the total pressure of the gaseous mixture.

Thus $K_{p/p°}$, defined as
$$\frac{\left(\dfrac{p_{NO_2}}{p°}\right)^2}{\dfrac{p_{N_2O_4}}{p°}},$$

$$= \frac{\left(\dfrac{2x}{(1+x)}\dfrac{p}{p^\circ}\right)^2}{\dfrac{(1-x)\,p}{(1+x)\,p^\circ}} = \frac{4x^2}{1-x^2} \cdot \frac{p}{p^\circ} .$$

Unlike the equilibrium constant K derived on p. 91, which is dependent on both temperature and pressure, K_{p/p° is independent of pressure at constant temperature provided the gases behave ideally.

Example

At 298 K, the value of $K_{p/p^\circ} = 0.100$ for the equilibrium

$$N_2O_4 \rightleftharpoons 2\ NO_2.$$

Calculate the fraction of N_2O_4 dissociated at this temperature when the pressure is 3.039×10^5 N m^{-2} i.e. $p = 3 \times p^\circ$.

$$\frac{4x^2}{1-x^2}\frac{p}{p^\circ} = \frac{12x^2}{1-x^2} = 0.100.$$

$$\therefore\ 12.1\,x^2 = 0.100$$

$$x = 0.091$$

It is seen that an increase in pressure decreases the extent of dissociation (cf. p. 97). In any gaseous reaction an increase in pressure favours that process which causes a decrease in the number of gas molecules. Thus in the reaction:

$$N_2 + 3\ H_2 \rightleftharpoons 2\ NH_3$$

the formation of ammonia is favoured by an increase in pressure, its dissociation by a decrease in pressure.

In a reaction such as:

$$H_2 + I_2 \rightleftharpoons 2\ HI$$

where there are equal numbers of gas molecules on the right and left sides of the equation, the equilibrium is not affected by change of pressure.

The use of equilibrium measurements in determining G° values

So far we have shown how conditions for equilibrium can be calculated if G° values are known for the chemicals involved in the reaction, but, conversely, equilibrium measurements can be used as one method of determining $G°$ values. The point can be illustrated by an example.

The density of PCl_5 vapour at 450 K and 1.013×10^5 N m^{-2} is 4.03 kg m^{-3}. But the molar mass M for PCl_5 is 0.2085 kg mol^{-1}. Thus the density is expected to be

$$\rho = \frac{Mp}{RT} \quad (\text{p.} \quad)$$

$$= \frac{0.2085 \text{ kg mol}^{-1} \times 1.013 \times 10^5 \text{ N m}^{-2}}{8.314 \text{ J K}^{-1} \text{ mol}^{-1} \times 450 \text{ K}}$$

$$= 5.65 \text{ kg m}^{-3}.$$

The reduction in density below the theoretical value for PCl_5 is explained by the dissociation

$$PCl_5 \text{ (g)} \rightleftharpoons PCl_3 \text{ (g)} + Cl_2 \text{ (g)}$$

If a fraction x of the PCl_5 is dissociated:

$$y_{Cl_2} = y_{PCl_3} = \frac{x}{1 + x}$$

a 1 $$y_{PCl_5} = \frac{1 - x}{1 + x}$$

Furthermore the number of gas molecules is increased in the ratio $(1 + x)/1$ by the dissociation. Thus, by Avogadro's law (p. 41),

$$1 + x = \frac{\text{theoretical density}}{\text{experimental density}} = \frac{5.65 \text{ kg m}^{-3}}{4.03 \text{ kg m}^{-3}}$$

$$= 1.402$$

and $$x = 0.402.$$

Thus $$K_{p/p°} = \frac{y_{PCl_3} \times y_{Cl_2}}{y_{PCl_5}} \times \frac{p}{p°}$$

For $p° = 1.013 \times 10^5$ N m^{-2}, $p = p°$,

and $$K_{p/p°} = \frac{x^2}{1 - x^2} = \frac{(0.402)^2}{1 - (0.402)^2} = 0.193 \text{ at } 450 \text{ K}$$

Thus $\Delta G°$ for the reaction at 450 K and 1.013×10^5 N m^{-2} is given by

$$\Delta G° = -RT \ln K_{p/p°}$$

$$= -8.314 \text{ J K}^{-1} \text{ mol}^{-1} \times 450 \text{ K} \times 2.303 \log_{10} 0.193$$

$$= +6.2 \text{ kJ mol}^{-1}.$$

For Cl_2 at 450 K and 1.013×10^5 N m^{-2} $G^\circ = -10.0$ kJ mol^{-1} and for PCl_3 under the same conditions $G^\circ = -450.1$ kJ mol^{-1}.

Since

$$\Delta G^\circ = G^\circ_{Cl_2} + G^\circ_{PCl_3} - G^\circ_{PCl_5}$$

$$\therefore \ G^\circ_{PCl_5} = G^\circ_{Cl_2} + G^\circ_{PCl_3} - \Delta G^\circ$$

$$= (-10.0 - 450.1 - 6.2) \text{ kJ mol}^{-1}$$

$$= -466.3 \text{ kJ mol}^{-1} \text{ at 450 K and } 1.013 \times 10^5 \text{ N m}^{-2}.$$

Heterogeneous equilibria

We shall discuss only one type of heterogeneous system, that of gas with solid. The general equation, Eqn. (36), still applies to a heterogeneous equilibrium but we use the convention that the activity of any solid constituent is unity. This is strictly true only if the total pressure is equal to p°, that for which the G° values apply, but the effect of pressure on the activity of a solid is generally small and can be neglected. Thus for a reaction such as

$$CaCO_3 \text{ (s)} \ \rightleftharpoons \ CaO \text{ (s)} + CO_2 \text{ (g)}$$

$$K_{p/p^\circ} = \frac{p_{CO_2}}{p^\circ},$$

assuming the gas to behave ideally. Thus the partial pressure of carbon dioxide at equilibrium is constant for a given temperature and is independent of the amounts of $CaCO_3$ and CaO present. Here an increase in total pressure favours the back reaction. Suppose $CaCO_3$, CaO and CO_2 are at equilibrium in an atmosphere of, say, nitrogen at 1.013×10^5 N m^{-2}. If the external pressure is doubled but p_{CO_2} remains the same the mole fraction and therefore the amount of CO_2 in the equilibrium mixture is halved.

Equilibrium in the reaction

$$3 \text{ Fe (s)} + 4 \text{ H}_2O \text{ (g)} \ \rightleftharpoons \ Fe_3O_4 \text{ (s)} + 4 \text{ H}_2 \text{ (g)}$$

is not affected by pressure, however, because the number of gas molecules is unaltered in the reaction (cf. p. 97).

Example

At 800 K and 1.013×10^5 N m^{-2}

$$G^\circ \text{ (H}_2\text{, g)} = -109 \text{ kJ mol}^{-1}$$
$$G^\circ \text{ (H}_2O\text{, g)} = -392 \text{ kJ mol}^{-1}$$

$$G^\circ \text{ (Fe, s)} \quad = - \quad 22 \text{ kJ mol}^{-1}$$
$$G^\circ \text{ (Fe}_3\text{O}_4, \text{ s)} = -1235 \text{ kJ mol}^{-1}.$$

Calculate the ratio $\dfrac{y\text{H}_2}{y\text{H}_2\text{O}}$ for the equilibrium:

$$3 \text{ Fe (s)} + 4 \text{ H}_2\text{O (g)} \rightleftharpoons 4 \text{ H}_2 \text{ (g)} + \text{Fe}_3\text{O}_4 \text{ (s) at 800 K.}$$

$$\Delta G^\circ \ = \ 4 \ (-109) \ + \ (-1235) - 3 \ (-22) - 4 \ (-392) \text{ kJ mol}^{-1}$$

$$= \ -37 \text{ kJ mol}^{-1}$$

$$\therefore \ \log_{10} K_{p/p^\circ} \ = \ \frac{-\Delta G}{2.303 \ RT} \ = \ \frac{+37 \ 000 \text{ J mol}^{-1}}{2.303 \times 8.314 \text{ J K}^{-1} \text{ mol}^{-1} \times 800 \text{ K}}$$

$$= \ 2.41$$

$$\therefore \ K_{p/p^\circ} \ = \ 260 \ = \left(\frac{y\text{H}_2}{y\text{H}_2\text{O}} \right)^4$$

$$\therefore \ \frac{y\text{H}_2}{y\text{H}_2\text{O}} \text{ at equilibrium at 800 K} \ = \ (260)^{\frac{1}{4}} = 4.0.$$

The effect of temperature on equilibrium

The equation:

$$\frac{d \ (\ln K_{p/p^\circ})}{dT} \ = \ \frac{\Delta H}{RT^2} \tag{48}$$

is the van't Hoff equation relating K_{p/p° to T. Integration between the limits T_1 and T_2 gives:

$$\ln \frac{K_2}{K_1} \ = \ \frac{\Delta H}{R} \left(\frac{1}{T_1} - \frac{1}{T_2} \right) \tag{49}$$

where K_1 is the value of K_{p/p° at T_1 and K_2 is the value at T_2. In the derivation of this equation ΔH is assumed to be constant within the temperature range. It can be seen from the equation that an increase in temperature favours the endothermic reaction.

Example

For the dissociation:

$$\text{F}_2 \rightleftharpoons 2 \text{ F}$$

at $1.013 \times 10^5 \text{ N m}^{-2}$, K_{p/p° at 825 K is exactly twice as large as K_{p/p° at 800 K. Calculate ΔH° for the reaction in this temperature range. Substituting in Eqn. (49):

$$2.303 \log_{10} 2 = \frac{\Delta H^\circ}{8.314 \text{ J K}^{-1} \text{ mol}^{-1}} \left(\frac{1}{800 \text{ K}} - \frac{1}{825 \text{ K}} \right)$$

$$\therefore \Delta H^\circ = \frac{2.303 \log_{10} 2 \times 8.314 \text{ J K}^{-1} \text{ mol}^{-1} \times 800 \text{ K} \times 825 \text{ K}}{25 \text{ K}}$$

$$= 152 \text{ kJ mol}^{-1}.$$

The Le Chatelier principle

The effects of temperature and pressure on equilibria are summarised qualitatively in this principle, which states that when the factors determining equilibrium are altered the system readjusts itself in such a way as to partly neutralise the change.

Let us consider the reaction:

$$N_2 \text{ (g)} + 3 H_2 \text{ (g)} \underset{\text{endothermic}}{\overset{\text{exothermic}}{\rightleftharpoons}} 2 NH_3 \text{ (g)}$$

An increase in pressure moves the equilibrium to the right by favouring the reduction in the number of gas molecules. An increase in temperature favours the endothermic process because some of the heat is thereby absorbed.

Examples VII

1. For the process:
$$Br_2 \text{ (l)} \longrightarrow Br_2 \text{ (g)}$$
at the boiling point, 332 K, $\Delta H^\circ = +30.6$ kJ mol^{-1}. Calculate ΔS° for the process at that temperature.

2. For the transition:

$$\text{White Tin} \longrightarrow \text{Grey Tin}$$

$\Delta H^\circ = -2.1$ kJ mol^{-1} (of Sn) at the transition temperature 291 K (i.e. the temperature at which the two forms of the element are in thermodynamic equilibrium). Calculate (a) ΔS° for the process (b) S° (grey tin) given that S° (white tin) = 51.5 J K^{-1} mol^{-1} (of Sn).

3. For the process:

$$K \text{ (s)} \longrightarrow K \text{ (l)}$$

ΔH° is +2.45 kJ mol^{-1} at the melting point, 336 K. Calculate ΔS° for the process at that temperature.

4. Calculate $\Delta S°$ for the reaction:

$$2 \text{ C (graphite)} + 3 \text{ H}_2 \text{ (g)} \longrightarrow \text{C}_2\text{H}_6 \text{ (g)}$$

at 298 K given the following standard entropies at that temperature. $S°$ (graphite) = 5.7 J K^{-1} mol^{-1}, $S°$ (H$_2$, g) = 130.5 J K^{-1} mol^{-1}, $S°$ (C$_2$H$_6$, g) = 228.5 J K^{-1} mol^{-1}.

5. Calculate $\Delta S°$ for the reaction:

$$\text{H}_2 \text{ (g)} + \text{Cl}_2 \text{ (g)} \longrightarrow 2 \text{ HCl (g)}$$

at 298 K and 1.013×10^5 N m^{-2} given the following standard entropies for those conditions. $S°$ (H$_2$, g) = 130.5 J K^{-1} mol^{-1}, $S°$ (Cl$_2$, g) = 223.0 J K^{-1} mol^{-1}, $S°$ (HCl, g) = 192.2 J K^{-1} mol^{-1}.

6. For the reaction:

$$2 \text{ HI (g)} \longrightarrow \text{H}_2 \text{ (g)} + \text{I}_2 \text{ (g)}$$

at 1000 K and 1.013×10^5 N m^{-2}, $\Delta H° = +10.0$ kJ mol^{-1} and $\Delta S° = -22.0$ J K^{-1} mol^{-1}. Calculate (a) $\Delta G°$ for the reaction at 1000 K, (b) $K_p/p°$ at 1000 K.

7. For the reaction:

$$\text{ZnO (s)} + \text{H}_2 \text{ (g)} \longrightarrow \text{Zn (g)} + \text{H}_2\text{O (g)}$$

at 1000 K and 1.013×10^5 N m^{-2}, $\Delta G° = +72.5$ kJ mol^{-1}. Calculate (a) $K_p/p°$ for the reaction (b) the vapour pressure of zinc in equilibrium with ZnO in the presence of hydrogen at 1000 K and 1.013×10^5 N m^{-2} total pressure.

8. Calculate $\Delta G°$ for the reaction:

$$2 \text{ NO (g)} + \text{O}_2 \text{ (g)} = \text{N}_2\text{O}_4 \text{ (g)}$$

at 298 K and 1.013×10^5 N m^{-2} given that, under these conditions, $H°$ (NO, g) = 90.5 kJ mol^{-1}, $H°$ (N$_2$O$_4$, g) = 9.7 kJ mol^{-1}, $S°$ (NO, g)= 210 J K^{-1} mol^{-1}, $S°$ (O$_2$, g) = 205 J K^{-1} mol^{-1}, $S°$ (N$_2$O$_4$, g) = 304 J K^{-1} mol^{-1}.

9. Calculate the mole fraction of the gas A in a mixture containing 0.35 mole of gas A, 0.41 mole of gas B and 0.72 mole of gas C only.

10. For an ideal gas A, $G° = 97.4$ kJ mol^{-1} at 298 K and 1.013×10^5 N m^{-2}. Calculate the contribution made by one mole of A to the total Gibbs function of a mixture of ideal gases, at a temperature of 298 K and a total pressure of 1.013×10^5 N m^{-2}, in which the mole fraction of A is 0.20.

11. In the dissociation of NOCl:

$$2 \text{ NOCl (g)} \rightleftharpoons 2 \text{ NO (g)} + \text{Cl}_2 \text{ (g)}$$

under certain conditions the degree of dissociation is x. Calculate K, the equilibrium constant in terms of mole fractions, as a function of x.

12. For the dissociation of tetracarbonyl nickel

$$\text{Ni(CO)}_4 \text{ (g)} \rightleftharpoons \text{Ni (s)} + 4 \text{ CO (g)}$$

calculate K_p/p° relative to a standard pressure $p^\circ = 1.013 \times 10^5$ N m^{-2} if the total pressure is the same as p° and the degree of dissociation is x.

13. Sulphur trioxide dissociates according to the reaction:

$$2 \text{ SO}_3 \text{ (g)} \rightleftharpoons 2 \text{ SO}_2 \text{ (g)} + \text{O}_2 \text{ (g)}$$

At 1000 K and 1.013×10^5 N m^{-2}, the density of the partly dissociated gas is 0.827 kg m^{-3}. $M \text{ (SO}_3) = 0.080$ kg mol^{-1}. Calculate the degree of dissociation of SO_3.

14. A mixture of N_2 and H_2 in which $y_{N_2} = 0.25$ is allowed to come to equilibrium. A fraction x of the nitrogen is converted to ammonia when equilibrium is reached:

$$N_2 + 3 H_2 \rightleftharpoons 2 NH_3.$$

If the pressure is p, express the value of K_p/p° as a function of of x, p and p°, the standard pressure.

15. For the dissociation:

$$\text{COCl}_2 \text{ (g)} \rightleftharpoons \text{CO (g)} + \text{Cl}_2 \text{ (g)}$$

at 373 K, K_p/p° relative to a standard pressure of 1.013×10^5 N m^{-2} is 6.7×10^{-9}. Calculate the partial pressure of CO in equilibrium at that temperature at a total pressure of 2.026×10^5 N m^{-2}.

16. The following ΔG° values refer to the reactions at 298 K and 1.013×10^5 N m^{-2}.

CO_2 (g) $+ 4 H_2$ (g)	$= CH_4$ (g) $+ 2 H_2O$ (g)	$\Delta G^\circ = -115$ kJ mol^{-1}	
$2 H_2$ (g) $+ O_2$ (g)	$= 2 H_2O$ (g)	$\Delta G^\circ = -456$ kJ mol^{-1}	
$2 C$ (s) $+ O_2$ (g)	$= 2 CO$ (g)	$\Delta G^\circ = -278$ kJ mol^{-1}	
C (s) $+ 2 H_2$ (g)	$= CH_4$ (g)	$\Delta G^\circ = - 52$ kJ mol^{-1}	

Calculate (a) ΔG° (b) K_p/p° at 298 K for the reaction:

$$CO_2 \text{ (g)} + H_2 \text{ (g)} = CO \text{ (g)} + H_2O \text{ (g)}$$

17. For the reaction:

$$N_2 \text{ (g)} + O_2 \text{ (g)} = 2 \text{ NO (g)}$$

the value of $K_p/p°$ is 1.10×10^{-3} at 2200 K and 3.60×10^{-3} at 2500 K. Use Eqn. (49) to calculate $\Delta H°$ for the reaction in that temperature range.

18. For the reaction:

$$Br_2 \text{ (g)} = 2 \text{ Br (g)}$$

$K_p/p° = 4.05 \times 10^{-4}$ at 1125 K and 7.20×10^{-3} at 1275 K. Calculate $\Delta H°$ for the reaction in that temperature range.

19. Choose from the following equilibria (i) those in which an increase in pressure favours the forward reaction, (ii) those in which pressure changes are without effect, (iii) those in which a decrease in pressure favours the forward reaction. Assume all the gases to behave ideally.

(a) $CO \text{ (g)} + H_2O \text{ (g)} \rightleftharpoons CO_2 \text{ (g)} + H_2 \text{ (g)}$
(b) $CaCO_3 \text{ (s)} \rightleftharpoons CaO \text{ (s)} + CO_2 \text{ (g)}$
(c) $CO \text{ (g)} + Cl_2 \text{ (g)} \rightleftharpoons COCl_2 \text{ (g)}$
(d) $C \text{ (s)} + CO_2 \text{ (g)} \rightleftharpoons 2 \text{ CO (g)}$
(e) $CO_2 \text{ (g)} + 4 H_2 \text{ (g)} \rightleftharpoons CH_4 \text{ (g)} + 2 H_2O \text{ (g)}$
(f) $NH_4Cl \text{ (s)} \rightleftharpoons NH_3 \text{ (g)} + HCl \text{ (g)}$
(g) $CO \text{ (g)} + 2 H_2 \text{ (g)} \rightleftharpoons CH_3OH \text{ (g)}$
(h) $CO \text{ (g)} + MgO \text{ (s)} \rightleftharpoons Mg \text{ (s)} + CO_2 \text{ (g)}$

20. Choose from the following equilibria (i) those in which an increase in temperature increases $K_p/p°$, (ii) those in which a decrease in temperature increases $K_p/p°$:

(a) $PCl_3 \text{ (l)} + Cl_2 \text{ (g)} = PCl_5 \text{ (s)}$ $\Delta H° = -139 \text{ kJ mol}^{-1}$
(b) $2 \text{ HCl (g)} = H_2 \text{ (g)} + Cl_2 \text{ (g)}$ $\Delta H° = +190 \text{ kJ mol}^{-1}$
(c) $3 \text{ Fe (s)} + 4 H_2O \text{ (g)} = Fe_3O_4 \text{ (s)} + 4 H_2 \text{ (g)}$ $\Delta H° = -146 \text{ kJ mol}^{-1}$.

Chapter 8

Chemical kinetics

A chemical equation expresses the stoichiometry of a reaction but tells us nothing about the rate or the mechanism of the process. Rates of reaction vary greatly; the rusting of iron, for example, is an extremely slow process; the reaction between sodium and water is a rapid one. Certain factors such as the nature, physical state or concentration of the reactants, temperature, the presence or absence of light or of catalysts, can greatly influence the speed of a particular reaction. The study of the reaction rates is called chemical kinetics.   ⟵

Of the factors which influence rate of reaction, we shall discuss two, concentration and temperature.

Methods of expressing reaction rates

The rates of chemical reactions are expressed by convention as rates of change of concentration. For a reaction:

$$\text{Reactants (R)} \longrightarrow \text{Products (P)}$$

the rate is defined as the rate of decrease of the concentration of R, i.e. $-d(c_R)/dt$ or as the rate of increase of the concentration of P, i.e. $d(c_P)/dt$. For the reaction represented by the equation

$$OH^- + C_2H_5Br = C_2H_5OH + Br^-$$

$$\frac{-d(c_{OH^-})}{dt} = \frac{-d(c_{C_2H_5Br})}{dt} = \frac{d(c_{C_2H_5OH})}{dt} = \frac{d(c_{Br^-})}{dt}$$

and any one of these rates of change of concentration can be considered as the reaction rate.

101

In the reaction:

$$H_2 + I_2 = 2\,HI$$

however, two moles of HI are produced for every mole of either H_2 or I_2 which reacts, thus:

$$\frac{d(c_{HI})}{dt} = -2\,\frac{d(c_{I_2})}{dt} = -2\,\frac{d(c_{H_2})}{dt}$$

It must therefore be emphasised that it is not sufficient to speak of a 'rate of reaction' without specifying the particular substance to which the rate refers.

Units

In the SI, concentrations are expressed in mol m^{-3} and time in seconds. Thus $\frac{d(c_R)}{dt}$ for a particular reactant is expressed in mol $m^{-3}\,s^{-1}$.

Order of reaction

For the reaction:

$$2\,N_2O_5\,(g) \longrightarrow 4\,NO_2\,(g) + O_2\,(g)$$

it is found by experiment that, at constant temperature,

$$\frac{-d(c_{N_2O_5})}{dt} = k(c_{N_2O_5})$$

where k is a constant called the rate constant for that particular temperature, its units being s^{-1}. The rate of decrease of $c_{N_2O_5}$ is proportional to $c_{N_2O_5}$. The reaction is said to be of first order.

For the reaction:

$$2\,NO_2\,(g) = 2\,NO\,(g) + O_2\,(g)$$

however, it is found that the rate at constant temperature is given by:

$$\frac{-d(c_{NO_2})}{dt} = k(c_{NO_2})^2.$$

The rate of decrease of c_{NO_2} is proportional to $(c_{NO_2})^2$ and the reaction is of second order. In general for a reaction:

$$R \longrightarrow P$$

for which $\dfrac{-d(c_R)}{dt} = k(c_R)^n$, the reaction is of nth order with respect to R.

For the reaction:

$$C_2H_4Br_2 + 2\ I^- = C_2H_4 + 2\ Br^- + I_2$$

the rate law is $\dfrac{-d(c_{I^-})}{dt} = k(c_{C_2H_4Br_2}) \times (c_{I^-})$; the reaction is of first order with respect to $C_2H_4Br_2$, of first order with respect to I^-, and of second order overall. In this equation k is a second-order rate constant expressed in $mol^{-1}\ m^3\ s^{-1}$. In general for a reaction

$$\text{Reactant A } + \text{ Reactant B} \longrightarrow \text{ Products}$$

in which the rate is defined as, say, $\dfrac{-d(c_A)}{dt}$, given by $\dfrac{-d(c_A)}{dt} = k(c_A)^a \times (c_B)^b$, the reaction is of ath order with respect to A, bth order with respect to B and of order $a + b$ overall.

Reactions of overall order greater than 2 are less common. Mention should be made of zero-order reactions, in which the reaction rate at a particular temperature is independent of the concentration of the reactant. The usual interpretation of zero-order kinetics is that the 'reactant', R, whose concentration is measured is not the true reactant, the latter, R', being a species produced from it, but at very low, constant concentration.

$$R \dashrightarrow R' \longrightarrow \text{ products}$$

$$\text{Rate } \alpha\ c_{R'} \text{ (i.e. constant)}$$

In other examples of zero-order reaction the products are removed as they are formed and the reactants are consequently not diluted as the reaction proceeds. If hydrogen and chlorine are kept over water in diffuse daylight, a slow zero-order reaction occurs. The quantities of H_2 and Cl_2 change but the concentrations do not, provided the total pressure is constant, because the HCl dissolves as it is formed.

Experimental determination of reaction rates

The usual procedure in determining a rate of reaction is to mix the reactants, in known concentrations, at the desired temperature, and subsequently to analyse the mixture for a

particular reactant at a series of time intervals. The analysis can be performed by titration (p. 178), as for example in
the reactions:

$$CH_3COOC_2H_5 + OH^- \text{ (aq)} = CH_3COO^- \text{ (aq)} + C_2H_5OH$$
$$H_2O_2 + 2 H_3O^+ + 2 I^- \text{ (aq)} = 4 H_2O + I_2,$$

by the measurement of a volume of gas evolved, as in the
reaction:

$$C_6H_5N_2^+ \text{ (aq)} + H_2O = C_6H_5OH + H^+ \text{ (aq)} + N_2 \text{ (g)},$$

by measuring the optical activity of the solution, as in the
reaction:

$$C_{12}H_{22}O_{11} \text{ (aq)} + H_2O = C_6H_{12}O_6 \text{ (aq)} + C_6H_{12}O_6 \text{ (aq)},$$

　　　sucrose　　　　　　　　　　glucose　　　　　fructuse

by precipitation, colorimetry, and many other methods.

Gas reactions can often be followed by the measurement
of pressure changes, as in the reaction:

$$CH_3CHO \text{ (g)} = CH_4 \text{ (g)} + CO \text{ (g)},$$

in which the number of molecules present increases as the
reaction proceeds, and the pressure exerted by a fixed volume
of gas also increases (p. 41). For fast reactions specialised
methods are required which we shall not discuss.

Rate equations and rate constants

(a) Zero-order reactions

Consider a reaction

$$R \longrightarrow \text{products}$$

for which $\dfrac{-d(c_R)}{dt} = k_0$ (the zero-order case). Integrating this
equation:

$$c_R = \text{an integration constant} - k_0t \qquad (50)$$

Thus if we plot the experimental value of $c_R/\text{mol m}^{-3}$
against ts^{-1}, where t is the time the reaction has been in pro-
gress, we obtain a straight line of slope $-k_0/\text{mol m}^{-3} \text{ s}^{-1}$ (Fig.
27).

(b) First-order reactions

Consider a reaction

$$R \longrightarrow \text{products}$$

for which $\dfrac{-d(c_R)}{dt} = k_1(c_R)$, the first-order case.

Suppose that $c_R = a$, originally

and $\qquad\qquad c_R = a - x$ after a time t.

Then

$$\frac{-d(c_R)}{dt} = \frac{-d(a-x)}{dt} = \frac{dx}{dt} \text{ (since } a \text{ is a constant)}$$

and

$$\frac{dx}{dt} = k_1(a-x)$$

Thus

$$k_1\,dt = \frac{dx}{(a-x)}$$

Integrating:

$$k_1\int_0^t dt = \int_0^x \frac{dx}{a-x}$$

because $x = 0$ when $t = 0$.

Thus

$$k_1 t = \ln \frac{a}{a-x} \qquad\qquad (51)$$

Since a is a constant concentration, it follows that a plot of $\longleftarrow$
$\ln \dfrac{a-x}{\text{mol m}^{-3}}$, i.e. $\ln \dfrac{c_R}{\text{mol m}^{-3}}$, against t s^{-1} gives a straight line
for a reaction of first order. (Fig. 28).

(c) Second-order reactions

Consider a reaction

$$R \longrightarrow \text{products,}$$

for which $\dfrac{-d(c_R)}{dt} = k_2(c_R)^2$ (the second-order case) in which c_R
$= a$ initially and $a - x$ after a time t.

Then $\qquad\qquad -\dfrac{d(a-x)}{dt} = k_2(a-x)^2$

i.e. $\qquad\qquad k_2 dt = -\dfrac{d(a-x)}{(a-x)^2}$

Integrating: $k_2 t = \dfrac{1}{a-x} +$ an integration constant

But since $x = 0$, when $t = 0$, the constant is $-\dfrac{1}{a}$

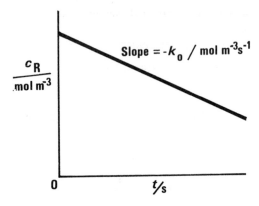

Fig. 27.   Zero-order graph.

$$\therefore k_2 t = \frac{1}{a-x} - \frac{1}{a} \tag{52}$$

Thus if we plot $\dfrac{\text{mol m}^{-3}}{c_R}$ against t s^{-1} we obtain a straight line
of slope k_2/mol^{-1} m^3 s^{-1}. (Fig. 29).

Example A

In dilute solution the reaction:

$$C_6H_5N_2Cl + H_2O = C_6H_5OH + H^+ + Cl^- + N_2$$

is found to be of first order. At 313 K the volume of nitrogen
liberated depends on time as follows:

time/ s	0	600	1200	1800	2400	3600	∞
Volume of N_2/ cm^3	0	6.4	11.8	14.0	16.2	19.0	21.8

Calculate the rate constant for the reaction.

Assuming nitrogen to be an ideal gas,

$$pV = nRT.$$

As p and T are constant in the experiment, and R is a universal
constant,

$$n \; \alpha \; V.$$

Thus the volume of nitrogen liberated at a certain time is
proportional to the amount of original benzene diazonium
chloride which has decomposed. Thus $c_{C_6H_5N_2Cl}$ at a particular
time is proportional to $V_\infty - V_t$ where V_∞ and V_t are the

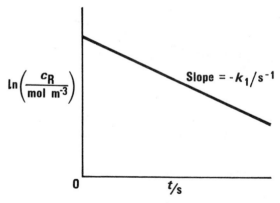

Fig. 28. First-order graph.

volumes of nitrogen at infinite time and at time t respectively.

Thus a plot of $\ln \dfrac{V_\infty - V_t}{cm^3}$ against t/s gives a line of slope $-k_1 s$.

time/s	$(V_\infty - V_t)/cm^3$	$\ln(\dfrac{V_\infty - V_t}{cm^3}) = 2.303 \log_{10} \dfrac{V_\infty - V_t}{cm^3}$
0	21.8	3.09
600	15.4	2.73
1200	10.0	2.30
1800	7.8	2.05
2400	5.6	1.75
3600	2.8	1.03

From these data, $k_1 = 5.72 \times 10^{-4}$ s^{-1} at 313 K.

Example B

The reaction between aniline and benzoyl chloride in benzene as solvent is of second order overall. The following data are for the reaction at 298 K.

t/s	$(c_{C_6H_5NH_2} = c_{C_6H_5COCl})/mol\ m^{-3}$
0	0.0200
300	0.0133
600	0.0100
1200	0.0067
1800	0.0051

Since $c_{C_6H_5NH_2} = c_{C_6H_5COCl}$ at any stage in the reaction the rate equation is that obtained in (c) on p. 105. Thus a plot of

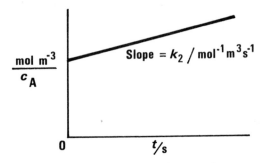

Fig. 29. Second-order graph.

$\dfrac{\text{mol m}^{-3}}{c_{C_6H_5NH_2}}$ against t/s should give a straight line of slope $k_2/\text{mol}^{-1}\ \text{m}^3\ \text{s}^{-1}$.

t	$\text{mol m}^{-3}/c_{C_6H_5NH_2}$
0	50.0
300	75.0
600	100.0
1200	149.0
1800	196.0

The line is found to have a slope of 8.3×10^{-2}. Thus $k_2 = 8.3 \times 10^{-2}\ \text{mol}^{-1}\ \text{m}^3\ \text{s}^{-1}$.

Determination of the order of a reaction

Consider a reaction between two species A and B:

$$A + B \longrightarrow \text{products.}$$

To find the order of the reaction with respect to A a series of experiments is carried out to determine the variation of c_A with time, the value of c_B being made so much larger than c_A that it remains effectively constant throughout the reaction. Trial graphs are drawn as in cases (a), (b) and (c) above. The order of the reaction with respect to A is indicated by the plot which gives the straight-line relation. A similar procedure is carried out with $c_A \gg c_B$ to find the order of reaction with respect to B. The effect of a large excess of one reactant, naturally, is to increase the reaction rate; thus the method previously described, while correct in principle, may become difficult in practice because some of the rates become too fast to measure by the methods described on p. 104.   In such cases

more specialised methods must be used which are outside the
scope of this book.

The half-life period

The half-life period, $t_{\frac{1}{2}}$, of a reaction:

$$R \longrightarrow products$$

is the time required for exactly one-half of R to be converted
into products.

Substituting in Eqn. (51) for the first-order case,
since

$$x = a/2 \text{ when } t = t_{\frac{1}{2}}$$

$$k_1 t_{\frac{1}{2}} = \ln \frac{a}{a - a/2} = \ln 2$$

$$\therefore t_{\frac{1}{2}} = \ln 2/k_1 = \frac{0.693}{k_1}$$

i.e. the value of $t_{\frac{1}{2}}$ is independent of the original concentration.
For zero-order and second-order reactions, $t_{\frac{1}{2}}$ depends on the
original concentration of R (Table XV).

Molecularity of a reaction

The reaction:

$$H_2 + I_2 = 2 \text{ HI,}$$

which is of first-order with respect to both hydrogen and iodine,
is believed to proceed as pictured below:

| H———— H | H - - - - -H | H | H |
| | | | |

Reactants   Transition state   Products

A diatomic molecule of hydrogen unites with a diatomic
molecule of iodine to form, momentarily, a transition-state
molecule containing four atoms. Reaction proceeds by the fission of the complex into two HI molecules. As two molecules
are required to produce the activated complex molecule, the
reaction is called bimolecular.

TABLE XV
Some Standard Kinetic Relations

Order of reaction	Rate equation	Simplest plot	Half-life
0	$\dfrac{-d(c_R)}{dt} = k_0$	$\dfrac{c_R}{\text{mol m}^{-3}}\Big/t\ \text{s}^{-1}$	$\dfrac{(c_R)\ \text{orig.}}{2\,k_0}$
1	$\dfrac{-d(c_R)}{dt} = k_1$	$\ln\dfrac{c_R}{\text{mol m}^{-3}}\Big/t\ \text{s}^{-1}$	$\dfrac{0.693}{k_1}$
2	$\dfrac{-d(c_R)}{dt} = k_2$	$\dfrac{\text{mol m}^{-3}}{c_R}\Big/t\ \text{s}^{-1}$	$\dfrac{1}{k_2\,(c_R)\ \text{orig.}}$

A <u>unimolecular</u> reaction is one in which a single molecule becomes <u>activated by</u> collision with another and subsequently decomposes to give products; only <u>one</u> molecule is required to form the transition state from which further reaction proceeds.

Reaction mechanisms

Our purpose in studying reaction rate and determining order of reaction is to gain insight into the way reactions proceed. An equation such as

$$S_2O_8^{2-}\ (aq)\ +\ 2\ I^-\ (aq)\ =\ 2\ SO_4^{2-}\ (aq)\ +\ I_2,$$

which represents the stoichiometry of the oxidation of iodide ions by peroxodisulphate ions, might lead us to expect (a) the formation of a transition complex from three ions (implying a termolecular process) and (b) a rate law:

$$\frac{d(c_{I_2})}{dt}\ =\ k\,(c_{I^-})^2\,(c_{S_2O_8^{2-}})$$

i.e. third-order kinetics overall. The experimental rate law is, however,

$$\frac{d(c_{I_2})}{dt}\ =\ k\,(c_{I^-})\,(c_{S_2O_8^{2-}});$$

the reaction is of <u>first order</u> with respect to both I^- and $S_2O_8^{2-}$ and therefore of <u>second order overall</u>. The following scheme represents a possible interpretation of the results.

(i) $S_2O_8^{2-} + I^- \longrightarrow (O_3SO \ldots OSO_3 \ldots I)^{3-} \longrightarrow SO_4^{2-} + SO_4^- + I$

                       transition state   <u>(slow)</u>

(ii) $SO_4^- + I^- \longrightarrow SO_4^{2-} + I$ <u>(fast)</u>

(iii) $I + I \longrightarrow I_2$ <u>(fast)</u>

A slow bimolecular process (i) is followed by two other bimolecular processes that are very rapid. The process (i) is called
the rate-determining process because the rate of the whole   ←
reaction depends on it alone.

It is clear from the previous paragraph that determination of order of reaction gives us guidance about the mechanism
of a rate-determining step alone; any scheme we devise for the
whole reaction must be somewhat tentative and intuitive. Of
course, cases often arise where the order of reaction is not an
integer because more than one of the reaction steps is measurably
slow. However, when steps in a reaction are identified they
nearly always prove to be unimolecular or bimolecular; termolecular reactions are extremely uncommon, the best known
being some of the reactions of nitric oxide, e.g.

$$2 \text{ NO} + O_2 = 2 \text{ NO}_2$$
$$2 \text{ NO} + Cl_2 = 2 \text{ NOCl}$$

which have overall third-order kinetics. Nitric oxide is, in
fact, a rather special case; the formation of a transition state
from three molecules, of which two are NO, is probably made
easy by the slight tendency of NO molecules to dimerise to N_2O_2
even in the absence of other species.

The influence of temperature on reaction rate

It is common for the rate of a reaction to increase approximately twofold for a rise in temperature of 10 K. If we consider the simplest case of a reaction in the gaseous phase, it is
evident (p. 48) that the number of collisions between molecules
is not greatly increased by such a small change in temperature,
hence the large increase in reaction rate must be explained in
some other way.

Figure 30 shows two curves representing the distribution
of molecular speeds in a gas at two temperatures differing by
10 K. It is evident that the proportion of molecules having a
speed greater than v_a at a particular moment, represented by
the shaded area, is increased by a large factor even though
the mean speed is increased very little.

Arrhenius (1889) argued that it was essential for two
colliding molecules to have a certain minimum amount of
energy, the activation energy, for them to be able to react. In   ←
a gas, the distribution of molecular speeds can be shown by the
theory of probability to be such that the fraction of molecules

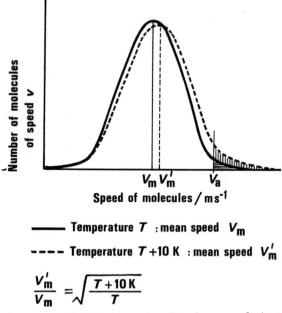

$$\frac{V'_m}{V_m} = \sqrt{\frac{T + 10\,K}{T}}$$

Fig. 30. Distribution of molecular speeds in a gas.

possessing an energy E at any instant is $\exp(-E/RT)$, where R is the gas constant and T the temperature.

Suppose that in a gas reaction only those molecules with energies greater than E are capable of entering into reaction, then E is the energy of activation for that reaction. The rate should be proportional to the number of effective collisions, i.e. the number of collisions multiplied by the fraction of molecules with sufficient energy to react. This argument led Arrhenius to propose the equation

$$k = A\exp(-E/RT) \qquad (53)$$

for the relation between the rate constant k and the activation energy E. The quantity A depends on the number of collisions and is approximately constant for a narrow range of temperature. It follows that if the overall rate constant of a reaction is k_1 at T_1 and k_2 at T_2, then

$$\ln\frac{k_1}{k_2} = \frac{-E}{R}\left(\frac{1}{T_1} - \frac{1}{T_2}\right) \qquad (54)$$

Thus if we plot ln (measure of k) against K$/T$ we obtain the Arrhenius plot, a straight line with a slope $\frac{-E}{R}$. An approximate value of E can be obtained from Eqn. (54) if values of k are known for only two temperatures.

Example

In view of the statement made on p. 111 it is of interest to calculate the energy of activation for a reaction whose rate constant doubles on raising the temperature from 298 K to 308 K.

Since, in this case $k_1 = \frac{1}{2}k_2$.

$$\therefore 2.303 \log_{10} \frac{1}{2} = \frac{-E}{R}\left(\frac{1}{298 \text{ K}} - \frac{1}{308 \text{ K}}\right)$$

$$= \frac{-E}{8.314 \text{ J mol}^{-1} \text{ K}^{-1}}\left(\frac{308 \text{ K} - 298 \text{ K}}{298 \text{ K} \times 308 \text{ K}}\right)$$

$$\text{and} -E = \frac{2.303 \times (-0.301) \times 298 \text{ K} \times 308 \text{ K} \times 8.314 \text{ J mol}^{-1} \text{ K}^{-1}}{(308 \text{ K} - 298 \text{ K})}$$

$$\therefore E = 5.29 \times 10^4 \text{ J mol}^{-1}$$

$$= 52.9 \text{ kJ mol}^{-1}.$$

Although reactions in solution are by their nature more complicated than gas reactions, it is usual to calculate experimental activation energies for them in exactly the same way as for gas reactions.

Examples VIII

1. For the reaction $2 \text{ NO} + \text{Cl}_2 = 2 \text{ NOCl}$ in the gas phase $\frac{-d(c_{\text{Cl}_2})}{dt} = k_3 \times (c_{\text{NO}})^2 \times (c_{\text{Cl}_2})$. Suppose that $c_{\text{NO}} = c_{\text{Cl}_2} = a$ at $t = 0$ and $c_{\text{NO}} = c_{\text{Cl}_2} = a - x$ at time t. Derive an equation for k_3 in terms of a, x and t. Suggest the best plot for a graphical determination of k_3.

2. The reaction

$$\text{C}_6\text{H}_5\text{N}_2\text{Cl} + \text{H}_2\text{O} = \text{C}_6\text{H}_5\text{OH} + \text{H}^+ + \text{Cl}^- + \text{N}_2$$

is of first order with respect to $\text{C}_6\text{H}_5\text{N}_2\text{Cl}$ in dilute aqueous solution. The amount of nitrogen evolved in the reaction varies with time at 323 K as follows:

time /s	360	540	720	1080	1440	1800	∞
Vol of N_2 /cm^3	19.3	26.0	32.6	41.3	46.5	50.4	58.3

Calculate (a) the first-order rate constant k_1, (b) $t_{\frac{1}{2}}$ for the reaction.

3. In a solution containing triethylamine and methyl iodide, both originally at a concentration of 20.0 mol m^{-3}, the following reaction occurred:

$$(C_2H_5)_3N + CH_3I = (C_2H_5)_3NCH_3^+ + I^-$$

The fraction of reactants converted to products varied with time as follows:

time /s	325	1295	1530	1975
fraction of reaction	0.314	0.649	0.688	0.737

Calculate the second-order rate constant for the reaction at that temperature.

4. The reaction:

$$SO_2Cl_2 \longrightarrow SO_2 + Cl_2$$

is a first-order gas reaction at 700 K with a rate constant 2.20 $\times 10^{-5}$ s^{-1}. What fraction of the SO_2Cl_2 will decompose if the gas is kept at 700 K for one hour?

5. A second-order reaction between two reactants, both originally at the same concentration, proceeds to the extent of 20% in 600 s. Calculate the time of half-reaction.

6. In concentrated sulphuric acid, formic acid reacts to give carbon monoxide gas.

$$H.COOH + H_2SO_4 = CO \ (g) + H_3O^+ + HSO_4^-.$$

The following results refer to the decomposition at 298 K:

time/s	0	50	100	150	200	∞
Vol of CO evolved /cm^3	0	11.6	20.2	26.1	30.4	41.5

Show by drawing a suitable graph that the reaction follows first-order kinetics and calculate the first-order rate constant.

7. Methyl acetate was hydrolysed in aqueous solution with an equimolar amount of alkali:

$$CH_3COOC_2H_5 + OH^- = CH_3COO^- + C_2H_5OH$$

Titrations with acid showed that the concentration of OH$^-$ ions in the solution at 298 K varied with time as follows:

time/s	0	300	600	900	1260	1500
c_{OH^-}/mol m^{-3}	10.0	6.34	4.64	3.63	2.88	2.54

Show by a suitable graph that the reaction is of second order overall, and calculate the rate constant at 298 K.

8. The hydrolysis of t-butyl chloride with a small quantity of water in anhydrous formic acid as solvent proceeds according to the equation:

$$(CH_3)_3CCl + H_2O \longrightarrow (CH_3)_3COH + H^+ + Cl^-$$

but follows a first-order rate law. Suggest how this result might be explained by a two-stage reaction mechanism.

9. The iodination of acetone by iodine in aqueous solution proceeds according to the equation:

$$CH_3COCH_3 + I_2 \longrightarrow CH_3COCH_2I + H^+ + I^-.$$

In the presence of a weak base, B, the rate law is:

$$\frac{-d(c_{I_2})}{dt} = k(c_{CH_3COCH_3}) \times (c_B).$$

Suggest a possible three-stage mechanism which would explain this result.

10. The following values were obtained for the second-order rate constant of the reaction between dimethyl aniline and methyl iodide in solution at different temperatures:

T/K	298	313	333	353
$10^2\,k$/mol^{-1} m^3 s^{-1}	7.16	21.0	77.2	238.0

By drawing a suitable graph, determine the experimental activation energy of the reaction.

11. The rate constant for the decomposition of acetonedicarboxylic acid in aqueous solution is found to vary with temperature as follows:

T/K	273	283	293	303	313	323	333
$10^5\,k$/s^{-1}	2.46	10.8	47.5	163	576	1850	5480

Draw a suitable graph and determine the experimental activation energy of the reaction.

12. For a first-order reaction $t_{\frac{1}{2}}$ is 3000 s at 313 K and 600 s at 333 K. Calculate the experimental activation energy of the reaction.

Chapter 9

Properties of solutions

A solution is a homogeneous, single-phase mixture of two or more components. In a two-component dilute solution a small amount of solute is present in a much larger amount of solvent. We shall be concerned here mainly with dilute solutions of non-volatile solutes in volatile solvents. Certain properties of such solutions, called colligative properties or osmotic properties, depend primarily on the number of discrete particles of solute in unit volume of solution and are independent of the nature of the particles.

Osmosis

If a dilute solution is separated from a more concentrated one by a semipermeable membrane, that is one which permits the passage of solvent but not solute molecules, the solvent will pass towards the more concentrated solution. Osmosis can be illustrated by stretching a membrane, such as pig's bladder, across a thistle funnel, inverting the latter, partly filled with an aqueous solution of, say, sucrose, in a beaker of water. The level of the liquid in the thistle funnel rises until the hydrostatic pressure acting on the membrane becomes sufficient to resist the passage of more solvent through it. The osmotic pressure of a solution is that pressure which must be applied to it to prevent solvent passing into it from pure solvent on the other side of a semipermeable membrane. Solutions having the same osmotic pressure are said to be isotonic.

The osmotic pressure, π, of a dilute solution is related to the amount of solute, n, and the total volume, V, by an equation analogous to the ideal gas equation (p. 41):

$$\pi = \frac{nRT}{V} \qquad (55)$$

116

Example

An aqueous solution containing 2.00 g sucrose in 100 cm^3 of water has an osmotic pressure of 1.40×10^5 N m^{-2} at 289 K. Calculate the relative molecular mass of the solute, M_r:

If the molar mass of the sucrose = M, then:

$$\frac{w}{M} = n = \frac{\pi V}{RT}$$

and $M = \dfrac{wRT}{\pi V}$

$$= \frac{2.00 \times 10^{-3} \text{ kg} \times 8.31 \text{ J K}^{-1} \text{ mol}^{-1} \times 289 \text{ K}}{1.40 \times 10^5 \text{ N m}^{-2} \times 1.00 \times 10^{-4} \text{ m}^3}$$

$$= 0.342 \text{ kg mol}^{-1}.$$

Since 1/12 of the mass of one mole of carbon-12 is exactly 10^{-3} kg mol^{-1}

$\therefore$ the relative molecular mass of sucrose, M_r, $= \dfrac{M}{1/12 \ M \ (C-12)}$

$$= \frac{0.342 \text{ kg mol}^{-1}}{1.0 \times 10^{-3} \text{ kg mol}^{-1}} = 342.$$

The value of M_r found in the experiment is the mean relative molecular mass of the particles of solute present in solution. The value found for the sucrose is for the monomer, $C_{12}H_{22}O_{11}$, the only solute species which is present in this case.

Lowering of vapour pressure

When a non-volatile solute is added to a volatile solvent the vapour pressure of the latter is lowered. Suppose a pure solvent has a vapour pressure, p, at a certain temperature, and that the addition of solute lowers the vapour pressure to p' at the same temperature, then the relative lowering of the vapour pressure is $(p-p')/p$. Raoult discovered by experiment that for dilute solutions of non-volatile solutes

$$\frac{p-p'}{p} = x_{solute} \quad \text{(Raoult's law)} \qquad (56)$$

where x_{solute}, the mole fraction of solute, is defined as

$\dfrac{n_{solute}}{n_{solute} + n_{solvent}}$ where solute and solvent refer to simple molecular species. For very dilute solutions this fraction becomes $n_{solute}/n_{solvent}$, thus

$$\frac{p - p'}{p} = \frac{n_{solute}}{n_{solvent}} \qquad (57)$$

An ideal solution obeys Raoult's law over a range of concentrations and temperatures (see also p. 124).

Example

Calculate the lowering of the vapour pressure of a 1.00% solution of sucrose ($M = 0.342$ kg mol^{-1}) at 371 K, at which temperature the vapour pressure of water is 9.42×10^4 N m^{-2}.

$$x_{C_{12}H_{22}O_{11}} = \frac{w_{C_{12}H_{22}O_{11}}}{M_{C_{12}H_{22}O_{11}}} \Big/ \frac{w_{H_2O}}{M_{H_2O}}$$

$$\therefore \frac{p - p'}{9.42 \times 10^4 \text{ N m}^{-2}} = \frac{1.00}{0.342 \text{ kg mol}^{-1}} \times \frac{0.018 \text{ kg mol}^{-1}}{(100 - 1)}$$

and $\qquad p - p' = 50.1$ N m^{-2}.

Methods are available for determining relative molecular masses in solution using the Raoult's law equation (Q. 6 p. 127), but, like measurements of osmotic pressure, they are not very reliable.

Elevation of boiling point

Figure 31 represents the variation of vapour pressure with temperature for (i) a pure solvent (ii) a solution of a non-volatile solute in that solvent. Since the vapour pressure of (i) is greater than that of (ii) at any temperature, the boiling point of the solution must be greater than that of the pure solvent, because the liquid boils when its vapour pressure is equal to the external pressure. The magnitude of the elevation of the boiling point, ΔT, depends on the molality , m, of the solute:

$$\Delta T = k_B m_{solute} = k_B \frac{w_{solute}}{M_{solute} \times W_{solvent}} \qquad (58)$$

where k_B is the ebullioscopic constant for the solvent. Thus the molar mass of a solute can be calculated from the elevation of boiling point, ΔT, produced when a known mass w is dissolved in a mass W of solvent:

$$M = \frac{k_B w}{W \Delta T} \qquad (59)$$

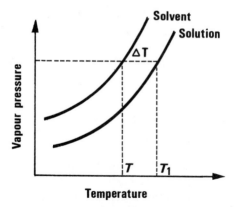

Fig. 31. Variation of vapour pressure with temperature for solvent and solution, illustrating the elevation of boiling point.

Example

1.40 g of a solute dissolved in 19.0 g water raised the boiling point of the latter by 0.50 K. Given that k_B for water = 0.52 K kg mol^{-1}, calculate M_r for the solute.

$$M = \frac{k_B w}{W \Delta T} = \frac{0.52 \text{ K kg mol}^{-1} \times 1.40 \times 10^{-3} \text{ kg}}{1.90 \times 10^{-2} \text{ kg} \times 0.50 \text{ K}}$$

$$= 0.077 \text{ kg mol}^{-1}$$

Thus $\quad M_r = \dfrac{0.077 \text{ kg mol}^{-1}}{1/12 \ M \ (C-12)} = 77.$

The experiment provides information on the actual relative molecular mass (molecular weight) of the molecules in solution, as shown in the following example:

A solution of 1.65 g of acetic acid in 100 g benzene boils 0.360 K above the boiling point of pure benzene, for which $k_B = 2.57$ K kg mol^{-1}.

$$\therefore \ M = \frac{k_B w}{W \Delta T} = \frac{2.57 \text{ K kg mol}^{-1} \times 1.65 \times 10^{-3} \text{ kg}}{1.00 \times 10^{-1} \text{ kg} \times 0.360 \text{ K}}$$

$$= 0.118 \text{ kg mol}^{-1}$$

$$\therefore \ M_r = \frac{0.118 \text{ kg mol}^{-1}}{1.00 \times 10^{-3} \text{ kg mol}^{-1}} = 118.$$

The value of M_r for $CH_3CO_2H = 60$, thus acetic acid must exist in benzene, under the conditions of the experiment, almost entirely as the dimer, which is known to be a hydrogen-bonded species of structural formula:

$$CH_3.C \underset{\displaystyle O - H ------ O}{\overset{\displaystyle O ------ H - O}{\big<}} \big> C.CH_3.$$

Depression of freezing point

The freezing point of a substance is the temperature at which the solid and the liquid have the same vapour pressure (Fig. 32). It therefore follows that a non-volatile solute, which lowers the vapour pressure of a solvent, also depresses its freezing point. The relation between the molality of the solute, m, and the depression of the freezing point, ΔT, is analogous to that for the elevation of boiling point (p. 118):

$$\Delta T = k_f m = k_f \frac{w}{MW} \tag{60}$$

where k_f is the cryoscopic constant for the solvent.

Example

A solution containing 1.00 g of X in 80.0 g of water has a freezing point 0.42 K lower than that of pure water. Calculate $M(X)$ given that k_f for water $= 1.86$ K kg mol^{-1}.

$$
\begin{aligned}
M &= \frac{k_f w}{W \Delta T} \\[6pt]
&= \frac{1.86 \text{ K kg mol}^{-1} \times 1.00 \times 10^{-3} \text{ kg}}{0.080 \text{ kg} \times 0.42 \text{ K}} \\[6pt]
&= 0.055 \text{ kg mol}^{-1}.
\end{aligned}
$$

For the determination of relative molecular masses in solution the freezing-point method is superior to measurement of osmotic pressure because temperatures can be measured more accurately than pressures; it is better than the boiling-point method because freezing points are affected far less than boiling points by small changes in atmospheric pressure. But, as in the other cases, the equation which we have given applies only to dilute solutions which behave ideally or nearly so.

Colligative properties of electrolytes

The depression of the freezing point produced by a dilute solution of an electrolyte is often greater, by a factor of two,

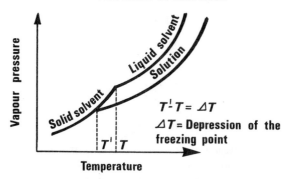

Fig. 32. Vapour pressure curves for solvent, solute and solid illustrating the depression of freezing point.

three or four, than that calculated from Eqn. (60). A dilute solution of $BaCl_2$ contains three times as many particles as a non-electrolyte of the same concentration because of the complete ionisation:

$$BaCl_2 \text{ (s)} + aq = Ba^{2+} \text{ (aq)} + 2 \text{ Cl}^- \text{ (aq)}$$

For a weak electrolyte capable of dissociating into n ions per molecule, the <u>degree of ionisation,</u> α, can be calculated from the <u>van't Hoff factor,</u> i, the number of times the freezing point depression or osmotic pressure exceeds that of a non-electrolyte of the same concentration:

$$i = 1 + \alpha (n - 1)$$

and

$$\alpha = \frac{i - 1}{n - 1} \tag{61}$$

Example

Aqueous $K_4Fe(CN)_6$ containing 1.00×10^2 mol m^{-3} at 290 K has an osmotic pressure of 6.87×10^5 N m^{-2}. Calculate the degree of ionisation:

$$K_4Fe(CN)_6 \rightleftharpoons 4 \text{ K}^+ + Fe(CN)_6^{4-}$$

For a non-electrolyte of concentration 1.00×10^2 mol m^{-3}

$$\pi = \frac{nRT}{V} \quad \text{(p. 116)}$$

$$= \frac{100 \text{ mol} \times 8.314 \text{ J K}^{-1} \text{ mol}^{-1} \times 290 \text{ K}}{1.00 \text{ m}^3}$$

$$= 2.41 \times 10^5 \text{ N m}^{-2}$$

$$\therefore \; i = \frac{6.87 \times 10^5 \text{ N m}^{-2}}{2.41 \times 10^5 \text{ N m}^{-2}} = 2.85$$

Thus
$$\alpha = \frac{2.85 - 1}{5 - 1} = 0.46.$$

Freezing points of non-ideal solutions

Even for solutions of non-electrolytes, the depression of the freezing point and other osmotic properties are not exactly proportional to concentration. In stronger solutions the solute molecules tend to interact with one another; consequently the osmotic effects they produce are somewhat smaller than expected — the solute becomes less 'active' relative to its concentration. The relative activity of a solute in solutions of different concentration can be determined by accurate determinations of freezing point depressions, but the mathematical treatment is outside the scope of this book. The concept of relative activity of a solute is however, an important one, particularly in connection with equilibria in liquids (p. 130).

The distribution law

A solute will distribute itself between two completely immiscible solvents in such a way that, at equilibrium, the ratio of concentration of solute in solvent A, c_A, to that in solvent B, c_B, is a constant, known as the distribution coefficient for a particular temperature:

$$\frac{c_A}{c_B} = K_D. \tag{62}$$

This is the Nernst distribution law. It is true only if the solute has the same molecular form in the two solvents. But if, for example, a solute S associates to form S_n molecules in one solvent but remains as a monomer in the other, the equilibrium becomes

$$nS \text{ (in solvent A)} \rightleftharpoons S_n \text{ (in solvent B)}$$

If concentrations are expressed as mol m^{-3} of monomer, then:

$$\left(\frac{c_A}{\text{mol m}^{-3}}\right)^n \Big/ \frac{c_B}{\text{mol m}^{-3}} = \text{a constant } K_{D'}.$$

The units are necessary, otherwise the ratio is not dimensionless.

Example A

For aniline distributed between benzene and water
$$\frac{c_{aniline} \text{ (in benzene)}}{c_{aniline} \text{ (in water)}} = 10.0 \text{ at 298 K.}$$ Calculate the weight of aniline extracted from a solution of 1.00 g aniline in 100 cm^3 of water by 20.0 cm^3 of benzene.

At equilibrium, if there is a weight x of aniline in the benzene,

$$\frac{x}{2.00 \times 10^{-5} \text{ m}^3} \Big/ \frac{1.00 \times 10^{-3} \text{ kg} - x}{1.00 \times 10^{-4} \text{ m}^3} = 10.0$$

$$\therefore \frac{x}{1.00 \times 10^{-3} \text{ kg} - x} = 10.0 \times \frac{2.00 \times 10^{-5} \text{ m}^3}{1.00 \times 10^{-4} \text{ m}^3}$$

$$= 2.00$$

$$\therefore x = 2.00 \times 10^{-3} \text{ kg} - 2x$$

$$3x = 2.00 \times 10^{-3} \text{ kg}$$

and $$x = 6.7 \times 10^{-4} \text{ kg} = 0.67 \text{ g.}$$

Example B

For benzoic acid distributed between benzene and water at 298 K the following results were obtained: c_A and c_B refer to the concentrations of C_6H_5COOH in water and benzene respectively:

| $10^{-1} c_A/\text{mol m}^{-3}$ | 4.01 | 3.18 | 2.31 | 1.77 | 1.37 |
| $10^{-2} c_B/\text{mol m}^{-3}$ | 11.0 | 6.86 | 3.51 | 2.07 | 1.27 |

Assuming the benzene to exist as C_6H_5COOH in water and as $(C_6H_5COOH)_n$ in benzene, calculate n.

Since $$\frac{(c_A/\text{mol m}^{-3})^n}{(c_B/\text{mol m}^{-3})} = \text{a constant,}$$

a plot of $\log_{10} \frac{c_B}{\text{mol m}^{-3}}$ against $\log_{10} \frac{c_A}{\text{mol m}^{-3}}$ gives a line of slope n.

| $\log_{10} \frac{c_A}{\text{mol m}^{-3}}$ | 1.604 | 1.502 | 1.363 | 1.248 | 1.137 |
| $\log_{10} \frac{c_B}{\text{mol m}^{-3}}$ | 3.041 | 2.836 | 2.545 | 2.316 | 2.104 |

The plot shows $n = 2$. Thus benzoic acid exists as $(C_6H_5COOH)_2$ in benzene.

The extraction of a solute from, say, aqueous solution using an immiscible organic solvent is an important technique in both organic and inorganic chemistry. Calculation shows (e.g. Q.16 and Q.17, p. 128) that several consecutive extractions with small volumes of solvent are always more efficient than a single extraction with the total volume of solvent.

Homogeneous mixtures of volatile liquids

Figure 33 represents the variation of vapour pressure with composition for a homogeneous mixture of 1,2 dibromoethane and 1,2 dibromopropane at constant temperature. The vapour pressure of the former, p_A, above the mixture, is related to the vapour pressure of pure 1,2 dibromoethane at the same temperature, $p°_A$, by

$$p_A = x_A p°_A \qquad (63)$$

where x_A is the mole fraction of 1,2 dibromoethane, $C_2H_4Br_2$. Similarly for 1,2 dibromopropane, $C_3H_6Br_2$:

$$p_B = x_B p°_B.$$

Hence the total vapour pressure exerted by the mixture is

$$p_A + p_B = x_A p°_A + x_B p°_B \qquad (64)$$

This is Raoult's law as it applies to an ideal mixture of volatile constituents. Ideal behaviour over a wide range of concentrations is quite rare because it implies that the interaction between molecules of A and B is identical with that between A and A and that between B and B.

Henry's law

Consider a very dilute solution of a volatile solute B in a solvent A. Molecules of B are present in such small numbers that any single molecule is surrounded entirely by molecules of A. In such circumstances the pressure exerted by B is given by:

$$p_B = kx_B \qquad (65)$$

where k is no longer equal to $p°_B$. This is Henry's law (1803). Applied to solutions of gases in liquids, it can be stated in the form: the solubility of a gas is directly proportional to the

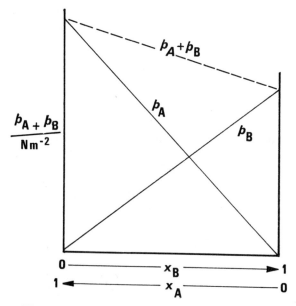

Fig. 33. Total vapour pressure over a mixture
of $C_2H_4Br_2$ (A) and $C_3H_6Br_2$ (B).

pressure at constant temperature. The law is true only if the
solute has the same molecular form in both phases. It is not
true for a solution of hydrogen chloride in water, for example,
because the gas contains covalent HCl molecules whereas the
aqueous phase contains H_3O^+ and Cl^- ions.

Since for a fixed amount of an ideal gas, pV = a constant
at constant temperature, the volume of a particular gas dissolved in unit volume of a particular liquid is a constant, independent of pressure, at a constant temperature. The Bunsen
absorption coefficient, α, is given by

$$\alpha = \frac{T_0 v}{TV} \tag{66}$$

where v = the volume of gas dissolved in a volume V of liquid
at a temperature T, and T_0 is a standard temperature, 273 K.
The absorption coefficient is clearly dimensionless. Some
values at 293 K for gases in water are:

$$O_2, \ \alpha \ = \ 0.0489,$$
$$N_2, \ \alpha \ = \ 0.0235,$$
$$H_2, \ \alpha \ = \ 0.0215.$$

In dry air at 293 K and 1.013×10^5 N m^{-2}, containing 21% by volume of O_2 and 79% by volume of N_2, the partial pressures (p. 44) are respectively

$$p_{O_2} = \frac{21}{100} \times 1.013 \times 10^5 \text{ N m}^{-2} = 2.13 \times 10^4 \text{ N m}^{-2}$$

and $p_{N_2} = \dfrac{79}{100} \times 1.013 \times 10^5$ N m$^{-2} = 8.00 \times 10^4$ N m^{-2}

Since, at constant temperature and volume for an ideal gas, n is proportional to p, the ratio $\dfrac{n_{O_2}}{n_{N_2}}$ in the air is $\dfrac{p_{O_2}}{p_{N_2}}$ and, in water which has come to equilibrium with air at 293 K,

$$\frac{n_{O_2}}{n_{N_2}} \text{ is } \frac{p_{O_2} \times \alpha_{O_2}}{p_{N_2} \times \alpha_{N_2}}$$

$$= \frac{2.13 \times 10^4 \text{ N m}^{-2} \times 0.0489}{8.00 \times 10^4 \text{ N m}^{-2} \times 0.0235} = 0.554.$$

Thus, the mole percentage of dissolved gas which is oxygen is:

$$\frac{0.554 \times 100}{1.554} = 35.6\% \ O_2.$$

If the solution is boiled, the gas which is evolved is thus almost twice as rich in oxygen as is ordinary air.

Examples IX

1. Calculate the osmotic pressure exerted at 285 K by an aqueous solution containing 1.50 mol m^{-3} of a non-electrolyte.

2. An aqueous solution containing 2.25 g of a non-electrolyte in 475 cm^3 exerts an osmotic pressure of 9.25×10^4 N m^{-2} at 290 K. Calculate the relative molecular mass of the solute.

3. An aqueous solution of an electrolyte MCl_3 containing 5.00 mol m^{-3} exerts an osmotic pressure of 4.26×10^4 N m^{-2} at 293 K. What is the degree of ionisation into M^{3+} and Cl^- ions at this concentration?

4. A solvent of molar mass 0.074 kg mol^{-1} has a vapour pressure of 1.06×10^4 N m^{-2} at 301 K. Calculate the lowering of vapour pressure caused by the addition of 19.3 g of a non-volatile solute of molar mass 0.235 kg mol^{-1} to 1.00 kg of solvent at 301 K.

5. At 298 K the vapour pressure of water is 3.16×10^3 N m^{-2}. Calculate the lowering of vapour pressure which results when 6.00 g urea ($M = 0.060$ kg mol^{-1}) is dissolved in 1.00 kg water at 298 K.

6. When a slow stream of a dry, unreactive gas is passed first through glass bulbs containing a solution of a non-volatile solute and then through bulbs containing pure solvent at the same temperature, both solution and solvent lose weight. If we assume the solvent in the vapour phase to behave as an ideal gas we can show that

$$\frac{\text{lowering of vapour pressure}}{\text{vapour pressure of solvent}} = \frac{\text{solvent evaporated from pure solvent}}{\text{total solvent evaporated}}$$

In such an experiment a solution containing 1.25 g of a non-volatile solute in 0.080 kg benzene ($M = 0.078$ kg mol^{-1}) lost 2.16 g and the pure benzene 0.0165 g. Calculate the molar mass of the solute.

7. A solution of 1.00 g of an organic solid in 50.0 cm^3 of ether boiled 0.63 K above the boiling point of pure ether, for which $k_B = 2.12$ K kg mol^{-1}. The density of ether is 714 kg m^{-3}. Calculate the molar mass of the solute.

8. The vapour pressure of pure ether is 5.90×10^4 N m^{-2} at 293 K. Calculate the vapour pressure exerted by the solution in Q. 7 above, at the same temperature. M (ether) $= 0.074$ kg mol^{-1}.

9. When 1.45 g of $CHCl_2.CO_2H$ is dissolved in 5.70×10^{-2} kg of CCl_4 the boiling point is raised by 0.520 K. k_B for CCl_4 is 5.25 K kg mol^{-1}. Explain this result.

10. For benzene $k_f = 5.12$ K kg mol^{-1}. Calculate the molar mass of a solute, 0.500 g of which, in solution, depresses the freezing point of 20.0 g benzene by 0.460 K.

11. A solution of 2.43 g sulphur in 100.0 g naphthalene has a freezing point 0.640 K below that of pure naphthalene, and a solution of 2.19 g of iodine in 100.0 g naphthalene has a freezing point 0.584 K below that of pure naphthalene. If iodine exists in naphthalene in the form of I_2 molecules what is the molecular formula of sulphur in the solvent?

12. A solution of acetamide in 0.100 kg of water began to freeze at 0.27 K below the freezing point of pure water. Calculate the

amount of ice which must separate for the temperature of the solution in equilibrium with it to fall a further 0.54 K.

13. Calculate the osmotic pressure exerted at 310 K by an aqueous solution which begins to freeze 0.56 K below the freezing point of pure water. k_f for water $= 1.86$ K kg mol^{-1}.

14. A certain weight of a substance dissolved in 0.100 kg benzene lowers the freezing point by 1.28 K. (k_f for benzene $= 5.12$ K kg mol^{-1}). The same weight dissolved in 0.100 kg water lowers the freezing point by 1.39 K. (k_f for water $= 1.86$ K kg mol^{-1}). If the substance exists in benzene in the simple molecular form, calculate the number of ions produced by one molecule in aqueous solution.

15. Experiments on the distribution of phenol between water and chloroform gave the following results, in which c_A and c_C refer to the concentrations of C_6H_5OH in water and chloroform respectively. Use a graphical method to find the value of n, assuming phenol to exist as C_6H_5OH in water and $(C_6H_5OH)_n$ in chloroform.

| c_A/mol m^{-3} | 1.00 | 1.73 | 2.63 | 4.64 | 5.81 |
| c_C/mol m^{-3} | 2.71 | 8.10 | 19.7 | 58.2 | 91.7 |

16. For a substance, X, the distribution coefficient between ether and water $= 3.00$, the substance being more soluble in ether. 100.0 cm^3 of an aqueous solution containing 10.0 g of X was shaken with (a) 50.0 cm^3 of ether (b) two successive quantities of 25.0 cm^3 of ether. Calculate the weight of X extracted in each case.

17. Calculate the amount of X which would be extracted under the conditions above if five successive portions of 10.0 cm^3 of ether were used.

18. At 343 K, $p°$ (chlorobenzene) $= 1.31 \times 10^4$ N m^{-2} and $p°$ (bromobenzene) $= 5.80 \times 10^3$ N m^{-2}. The liquids form an ideal mixture. Calculate the vapour pressure of a solution containing 0.20 mol C_6H_5Br and 0.80 mol C_6H_5Cl at 343 K.

19. The normal boiling point of n-hexane is 342 K. At that temperature $p°$ (n-heptane) $= 3.94 \times 10^4$ N m^{-2}. The liquids form an ideal solution. Calculate the total vapour pressure of a solution which contains 21.5 g n-hexane ($M_r = 86.0$) in 60.0 g n-heptane ($M_r = 100.0$) at 342 K.

20. Use the Bunsen absorption coefficients on p. 125 to calculate the mole percentage of oxygen in the dissolved gas when

water is equilibrated with a gas containing H_2 and O_2 in the ratio 2:1 by volume at 293 K.

Chapter 10

Ionic equilibria in aqueous solution

Solubility

The equilibrium between an ionic solid, e.g. MX, and its saturated aqueous solution can be represented as

$$MX \ (s) + aq \rightleftharpoons M^{z+}(aq) + X^{z-} \ (aq)$$

The solubility of the solid at a specified temperature is the mass which dissolves divided by the mass of the solvent when equilibrium is reached in the presence of an excess of the solute. Solubility is therefore a dimensionless quantity. The solubility divided by the molar mass is the molality of the saturated solution, m_{sat} (pp. 118 and 207).

Example

The mass of copper(I) iodide which dissolves in 50 g of pure water when equilibrium is reached with the solid salt at 298 K is 15.4 μg. The solubility is therefore given by

$$s = \frac{15.4 \times 10^{-6} \text{ g}}{50 \text{ g}} = 3.08 \times 10^{-7}.$$

Since the molar mass of CuI, $M(\text{CuI})$, = 0.191 kg mol^{-1}, the molality of the saturated aqueous solution of CuI at 298 K is

$$m_{sat} = \frac{s}{M} = \frac{3.08 \times 10^{-7}}{0.191 \text{ kg mol}^{-1}}$$

$$= 1.61 \times 10^{-6} \text{ mol kg}^{-1}.$$

Ionic activity

The relative activity of an ion in solution, a_i, is a dimensionless quantity:

$$a_i = \frac{m_i \gamma_i}{m^\circ} \qquad (67)$$

where m_i is the molality of the ion in solution, m° is a standard molality taken as 1 mol kg^{-1} and γ_i is the <u>activity coefficient</u> of the ion. For dilute, aqueous solutions ($m < 10^{-3}$ mol kg^{-1}) of 1:1 electrolytes at 298 K the value of γ_i can be calculated, within 0.1%, from the equation

$$\log_{10} \gamma_i \cdot = -0.51 \text{ kg}^{\frac{1}{2}} \text{ mol}^{-\frac{1}{2}} \sqrt{m_i} \qquad (68)$$

Thus for a saturated solution of CuI in which, for both Cu$^+$ and I$^-$ ions, $m_i = 1.61 \times 10^{-6}$ mol kg^{-1}

$$\log_{10} \gamma_i = -0.51 \text{ kg}^{\frac{1}{2}} \text{ mol}^{-\frac{1}{2}} (1.61 \times 10^{-6} \text{ mol kg}^{-1})^{\frac{1}{2}}$$

$$= -6.47 \times 10^{-4}$$

$$= -0.000647$$

and $\quad \gamma_i = \bar{1}.9994$ (to the fourth place)

$$= 0.9987.$$

This is so close to unity that we can write Eqn. (67) in the form

$$a_i = \frac{m_i}{\text{mol kg}^{-1}} \qquad (69)$$

In this book we shall use this equation for relative activity for solutions of 1:1 electrolytes in which $m_i < 10^{-3}$ mol kg^{-1} and for other electrolytes in which $m_i < 10^{-4}$ mol kg^{-1}.

Thus for saturated aqueous CuI

$$a \text{ (Cu}^+\text{)} = a \text{ (I}^-\text{)} = \frac{1.61 \times 10^{-6} \text{ mol kg}^{-1}}{\text{mol kg}^{-1}}$$

$$= 1.61 \times 10^{-6}$$

Solubility product

Consider the equilibrium between an excess of solid CuI and its ions in aqueous solution:

$$\text{CuI (s)} + \text{aq} = \text{Cu}^+ \text{(aq)} + \text{I}^- \text{(aq)}.$$

It is convenient to take the activity of a pure compound as unity, thus $a_{CuI} = a_{H_2O} = 1$; the water contains so little dissolved salt that it is effectively pure water. The equilibrium constant therefore becomes

$$\frac{a_{Cu^+} \times a_{I^-}}{1 \times 1} = a_{Cu^+} \times a_{I^-} = K_S.$$

→ The constant K_S is called the <u>solubility product</u> for CuI at the specified temperature. From the previous paragraph

$$K_S \text{ for CuI at 298 K} = (1.61 \times 10^{-6})^2 = 2.60 \times 10^{-12}.$$

In general, for an ionic solid MX_n which ionises

$$MX_n + aq = M^{n+} (aq) + nX^- (aq),$$

$$K_S = a_{M^{n+}} \times (a_{X^-})^n \qquad (70)$$

where $a_{M^{n+}}$ and a_{X^-} represent the activities of these ions in a solution saturated with MX_n at the specified temperature.

For $Fe(OH)_3$ the solubility in pure water at 298 K is 4.82 $\times 10^{-11}$. The molar mass is 0.107 kg mol^{-1}.

Thus $\qquad m_{sat} = \dfrac{4.82 \times 10^{-11}}{0.107 \text{ kg mol}^{-1}} = 4.50 \times 10^{-10} \text{ mol kg}^{-1}$

Thus $\qquad a_{Fe^{3+}} = 4.50 \times 10^{-10}$

and $\qquad a_{OH^-} = 3 \times 4.50 \times 10^{-10}$,

(since three moles OH$^-$ are produced by every mole of $Fe(OH)_3$ which dissolves).

Thus K_S for $Fe(OH)_3$ at 298 K

$$= a_{Fe^{3+}} \times (a_{OH^-})^3 \text{ for the saturated solution}$$

$$= 4.50 \times 10^{-10} \times (3 \times 4.50 \times 10^{-10})^3$$

$$= 1.11 \times 10^{-36}.$$

The solubility product principle

Since the value of K_S is a constant for a given salt MX at a specified temperature, it follows that the solubility of MX is lowered by the presence of an excess of either the cation M^{z+} or the anion X^{z-} in solution. Furthermore, the addition of a common ion to a saturated solution of a salt causes some of that salt to be precipitated. For example, when HCl gas is passed into saturated aqueous NaCl, some NaCl is precipitated because the additional Cl$^-$ ions produced by the reaction:

$$HCl + aq = H^+ (aq) + Cl^- (aq)$$

cause the solubility product $K_s = a_{Na^+} \times a_{Cl^-}$ to be exceeded. Conversely, the removal from solution of one of the ions in equilibrium with the solid causes more of the solid to dissolve. Thus if HCl is added to an equilibrium mixture of solid calcium oxalate and its saturated solution:

$$CaC_2O_4 + aq \rightleftharpoons Ca^{2+} (aq) + C_2O_4^{2-} (aq),$$

some $C_2O_4^{2-}$ ions are removed by the reaction

$$2 H^+ (aq) + C_2O_4^{2-} (aq) \longrightarrow H_2C_2O_4 (aq)$$

because oxalic acid is a weak electrolyte (p. 165). As a result, more calcium oxalate dissolves to restore the product $a_{Ca^{2+}} \times a_{C_2O_4^{2-}}$ to the K_s value of the salt at that temperature.

Example

Calculate the solubility of $PbCrO_4$ in an aqueous solution containing 1.00×10^{-3} mol kg^{-1} of CrO_4^{2-} ion. K_s for $PbCrO_4$ $= 1.80 \times 10^{-14}$; $M (PbCrO_4) = 0.323$ kg mol^{-1}.

Since the molality of CrO_4^{2-} in the solution

$$= 1.00 \times 10^{-3} \text{ mol kg}^{-1},$$

$$a_{CrO_4^{2-}} = 1.00 \times 10^{-3} \text{ (p. 131)}.$$

But $$K_s = a_{Pb^{2+}} \times a_{CrO_4^{2-}} = 1.80 \times 10^{-14}$$

$$\therefore a_{Pb^{2+}} = \frac{1.80 \times 10^{-14}}{1.00 \times 10^{-3}} = 1.80 \times 10^{-11}$$

Thus the saturated solution contains 1.80×10^{-11} mol kg^{-1} and the solubility of $PbCrO_4$ in the presence of 1.00×10^{-3} mol kg^{-1} of CrO_4^{2-} ion is:

$$1.80 \times 10^{-11} \text{ mol kg}^{-1} \times 0.323 \text{ kg mol}^{-1}$$

$$= 5.82 \times 10^{-12}.$$

Weak electrolytes

The equilibrium between ions in solution and a solid salt is an example of <u>heterogeneous</u> equilibrium between a liquid phase and a solid phase. But <u>homogeneous</u> equilibria in the liquid phase can exist between molecules and ions which are produced from them. We shall limit our discussion to the production of ions by acids and bases in aqueous solution. An <u>acid</u> was defined independently by Lowry and by Brønsted in the same year, 1920, as a species which has a tendency to lose a proton. When an acid such as acetic acid is dissolved in water an equilibrium is set up:

$$CH_3COOH + H_2O \rightleftharpoons CH_3COO^- + H_3O^+.$$

The reaction is called protolysis because it involves the transfer of a proton from the acid, in this case acetic acid, to a water molecule, thereby converting the latter into the hydroxonium ion, H_3O^+.

Lowry and Brønsted defined a base as a species tending to gain a proton; thus the water molecule in the protolysis reaction above is acting as a base. In the reverse reaction, the acetate ion acts as a base in accepting a proton from a hydroxonium ion. The acetate ion is called the conjugate base of the acetic acid molecule, the hydroxonium ion is the conjugate acid of the water molecule. The H_3O^+ ion is a cationic acid; another example is the ammonium ion, NH_4^+, which reacts with water to produce the base ammonia:

$$NH_4^+ + H_2O \rightleftharpoons NH_3 + H_3O^+.$$

An acid and the base produced from it by the loss of one proton are referred to as a conjugate acid-base pair. Anionic acids are also known; an example is the bisulphate ion HSO_4^- which has as its conjugate base the sulphate ion, SO_4^{2-},

$$HSO_4^- \quad + \quad H_2O \rightleftharpoons SO_4^{2-} \quad + \quad H_3O^+$$

(Acid A)   (Base B) $\begin{pmatrix} \text{Conjugate} \\ \text{base of A} \end{pmatrix}$ $\begin{pmatrix} \text{Conjugate} \\ \text{acid of B} \end{pmatrix}$

Acid dissociation constant

The equilibrium constant for the protolysis

$$CH_3COOH + H_2O \rightleftharpoons CH_3COO^- + H_3O^+$$

is given by

$$K = \frac{a_{CH_3COO^-} \times a_{H_3O^+}}{a_{CH_3COOH} \times a_{H_2O}}$$

For dilute solutions it is sufficient to take $a_{H_2O} = 1$ and the relative activities of the other species as equal to their molalities divided by 1 mol kg^{-1}. The dissociation constant, K_a, for the acid is defined by the equation

$$K_a = \frac{a_{\text{conjugate base}} \times a_{H_3O^+}}{a_{\text{acid}}} \tag{71}$$

i.e.
$$\frac{a_{CH_3COO^-} \times a_{H_3O^+}}{a_{CH_3COOH}} \quad \text{for acetic acid.}$$

The value of K_a for acetic acid is 1.75×10^{-5}.

Consider a dilute aqueous solution of acetic acid in which a fraction α of the molecules is converted to acetate ions when equilibrium is achieved. The fraction α is known as the degree of dissociation of the acid. Then for a solution of molality m relative to CH_3COOH, the equilibrium activities of acetate ion, hydroxonium ion and acid are given by:

$$a_{CH_3COO^-} = \frac{\alpha m}{\text{mol kg}^{-1}} \quad \text{(Eqn. 69, p. 131)}$$

$$a_{H_3O^+} = \frac{\alpha m}{\text{mol kg}^{-1}}$$

$$a_{CH_3COOH} = \frac{(1-\alpha)m}{\text{mol kg}^{-1}}$$

Thus
$$K_a = \frac{\alpha^2 m}{(1-\alpha) \text{ mol kg}^{-1}} \qquad (72)$$

When the degree of dissociation is small, the relation between K_a and α is given to a reasonable approximation by equating $(1-\alpha)$ to unity; thus:

$$K_a = \frac{\alpha^2 m}{\text{mol kg}^{-1}} \qquad (73)$$

Thus for a solution of acetic acid containing 1.00×10^{-2} mol kg^{-1}, the value of α is obtained from

$$\frac{\alpha^2 \times 1.00 \times 10^{-2} \text{ mol kg}^{-1}}{\text{mol kg}^{-1}} = 1.75 \times 10^{-5}$$

whence
$$\alpha^2 = 1.75 \times 10^{-3}$$
and
$$\alpha = 4.2 \times 10^{-2}.$$

The acid dissociation exponent, pK_a

It has been found useful, particularly for the tabulation of data on acid dissociation, to introduce a dissociation exponent, pK_a, defined for a particular acid by

$$pK_a = -\log_{10} K_a \qquad (74)$$

Thus for acetic acid, for which $K_a = 1.75 \times 10^{-5}$ at 298 K,

$$pK_a = -\log_{10}(1.75 \times 10^{-5})$$
$$= -(\bar{5}.24)$$
$$= +5.00 - 0.24 = 4.76.$$

Table XVI contains the pK_a values for some acids. The following points are of interest:

(i) Very strong acids, which ionise virtually completely, have large negative pK_a values. The more positive the pK_a value, the weaker the acid.

(ii) Acid strengths increase in a series such as:

$$CH_3.COOH < CH_2Cl.COOH < CHCl_2.COOH < CCl_3.COOH$$

The electronegative Cl atoms tend to draw electrons away from the oxygen of the OH bond, thus stablising the anion, i.e. the conjugate base:

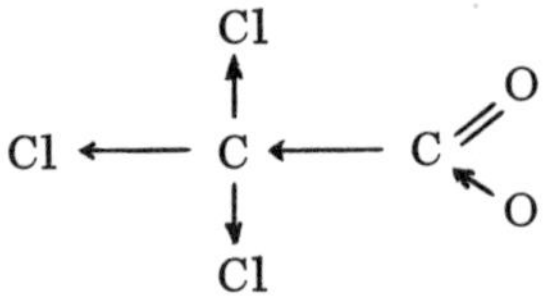

The withdrawal of electrons from one part of a molecule by a strongly electronegative group in another part is called an inductive effect.

(iii) Acid strengths decrease in the series:

$$H_3PO_4 > H_2PO_4^- > HPO_4^{2-}.$$

The tendency of a species to lose a proton decreases as successive protons are removed.

The autoprotolysis constant for water, K_W

Measurements of the electric conductance of pure water show that for the autoprotolysis:

$$H_2O + H_2O \rightleftharpoons H_3O^+ + OH^-$$

(i.e., the transfer of a proton from one H_2O molecule to a neighbouring one)

$$K_W = a_{H_3O^+} \times a_{OH^-} \tag{75}$$
$$= 1.00 \times 10^{-14} \text{ at } 298 \text{ K}$$

TABLE XVI
pK_a Values for Some Acids

HCl	ca -7	HCN	9.15
HBr	ca -9	HNO$_2$	3.4
HClO$_4$	ca -8	HSO$_4^-$	2.0
H$_2$SO$_4$	ca -3	NH$_4^+$	9.25
CH$_3$COOH	4.76	C$_2$H$_5$NH$_3^+$	10.25
CH$_2$Cl.COOH	2.81	C$_6$H$_5$NH$_3^+$	4.66
CHCl$_2$.COOH	1.3	HPO$_4^{2-}$	12.3
CCl$_3$.COOH	0.7	H$_2$PO$_4^-$	7.2
C$_2$H$_5$.COOH	4.87	H$_3$PO$_4$	2.12
C$_6$H$_5$.COOH	4.17		

The strengths of bases

When ethylamine dissolves in water the equilibrium

$$C_2H_5NH_2 + H_2O \rightleftharpoons C_2H_5NH_3^+ + OH^-$$

is set up. It is convenient to use a base dissociation constant K_b:

$$K_b = \frac{a_{C_2H_5NH_3^+} \times a_{OH^-}}{a_{C_2H_5NH_2}}$$

i.e.,

$$K_b = \frac{a_{\text{conjugate acid}} \times a_{OH^-}}{a_{\text{base}}} \tag{76}$$

K_b for a base is related to K_a for its conjugate acid and to K_W as follows:

$$K_b \times K_a = \frac{a_{\text{conjugate acid}} \times a_{OH^-}}{a_{\text{base}}} \times \frac{a_{\text{base}} \times a_{H_3O^+}}{a_{\text{conjugate acid}}}$$

$$= a_{OH^-} \times a_{H_3O^+} = K_W.$$

Thus

$$K_b = \frac{K_W}{K_a \text{ for the conjugate acid}} \tag{77}$$

If we define pK_b as $-\log_{10} K_b$, it follows that

$$pK_b = 14.0 - pK_a \text{ (conjugate acid)}, \tag{78}$$

since $\log_{10} K_W = -14.0$.

Thus, for ethylamine,

$$pK_b = 14.0 - 10.25 \quad \text{(Table XVI)}$$

$$= 3.75.$$

Similarly, for the acetate ion:

$$pK_b = 14.0 - 4.76$$
$$= 9.24.$$

and for the chloride ion

$$pK_b = 14.0 - (-7.0)$$
$$= 21.0.$$

The chloride ion is evidently a very weak base indeed. It is evident that the conjugate base of a strong acid (e.g. HCl) is a very weak base whereas the conjugate base of a weak acid such as $C_2H_5NH_3^+$ is a strong base.

In recent years it has become the custom not to tabulate pK_b values for bases in the reference literature but to give only the pK_a values of the conjugate acids. The reader can always calculate pK_b by the use of Eqn. (77), however.

Example

For the acid $C_6H_5NH_3^+$, the anilinium ion, $pK_a = 4.66$. Calculate pK_b for aniline, $C_6H_5NH_2$

$$pK_b = 14.0 - 4.66$$
$$= 9.34$$

The hydrogen ion exponent, pH

For all aqueous solutions at 298 K the product

$$a_{H_3O^+} \times a_{OH^-} = 1.00 \times 10^{-14}.$$

Sorenson (1909) introduced the concept of the hydrogen ion exponent, pH, of an aqueous solution, which he defined by the equation:

$$pH = -\log_{10} a_{H_3O^+} \text{ for that solution} \qquad (79)$$

Because of the difficulty of measuring relative activities precisely, pH scales are now based upon certain standard solutions, but we shall use the Sorenson definition in this book.

Since in an exactly neutral solution at 298 K,

$$a_{H_3O^+} = a_{OH^-} = (1.00 \times 10^{-14})^{\frac{1}{2}}.$$
$$\therefore a_{H_3O^+} = 1.00 \times 10^{-7}$$

and
$$pH = 7.0.$$

An aqueous solution in which pH < 7.0 is an <u>acidic</u> solution, one
in which pH > 7.0 is an <u>alkaline</u> solution.

Example A

Calculate the pH of an aqueous solution of acetic acid (K_a = 1.75 × 10^{-5}) containing 1.00 × 10^{-2} mol kg^{-1} CH$_3$COOH.

For the equilibrium:

$$CH_3COOH + H_2O \rightleftharpoons CH_3COO^- + H_3O^+$$

$$K_a = \frac{a_{CH_3COO^-} \times a_{H_3O^+}}{a_{CH_3COOH}} = 1.75 \times 10^{-5}$$

Since, from the equation for the reaction, $a_{CH_3COO^-} = a_{H_3O^+}$ and since

$$a_{CH_3COOH} = \frac{1.00 \times 10^{-2} \text{ mol kg}^{-1}}{\text{mol kg}^{-1}}$$

$$= 1.00 \times 10^{-2},$$

$$\therefore \quad \frac{(a_{H_3O^+})^2}{1.00 \times 10^{-2}} = 1.75 \times 10^{-5}$$

$$\therefore \quad a_{H_3O^+} = (1.75 \times 10^{-7})^{\frac{1}{2}}$$

$$= 4.18 \times 10^{-4}$$

and

$$pH = -\log_{10}(4.18 \times 10^{-4})$$

$$= -(\bar{4}.62)$$

$$= +4.0 - 0.62$$

$$= 3.38.$$

Example B

Calculate the pH of an aqueous solution of ethylamine (pK_b = 3.75) containing 1.00 × 10^{-3} mol kg^{-1} of C$_2$H$_5$NH$_2$ at 298 K

$$C_2H_5NH_2 + H_2O \rightleftharpoons C_2H_5NH_3^+ + OH^-$$

$$K_b = 10^{-3.75} = 1.78 \times 10^{-4}$$

$$(\text{Note: } -3.75 = \bar{4}.25)$$

$$\frac{(a_{OH^-})^2}{1.00 \times 10^{-3}} = 1.78 \times 10^{-4}$$

$$\therefore \quad a_{OH^-} = (1.78 \times 10^{-7})^{\frac{1}{2}}$$

$$= 4.20 \times 10^{-4}.$$

Since $\quad\quad a_{OH^-} \times a_{H_3O^+} = 1.00 \times 10^{-14}$ at 298 K.

$$\therefore a_{H_3O^+} = \frac{1.00 \times 10^{-14}}{4.20 \times 10^{-4}}$$

and $\quad\quad\quad\quad\quad\quad pH = -(\overline{11}.38)$

$$= 10.62$$

Hydrolysis

When a salt such as sodium acetate is dissolved in water the solution is found to have an alkaline reaction. The solid salt is ionic, consisting of Na^+ ions and CH_3COO^- ions; its reaction with water can be represented as:

$$Na^+ + CH_3COO^- + H_2O \rightleftharpoons Na^+ + CH_3COOH + OH^-$$

The Na^+ ion exists in solution as an aquated ion $Na(H_2O)_n^+$, but this species has a very low tendency to lose a proton (i.e. it is a very weak acid) and is considered to take no part in the reaction. The acetate ion, however, tends to capture a proton from a water molecule; thus the equation above can be simplified to:

$$CH_3COO^- + H_2O = CH_3COOH + OH^-$$

The equilibrium constant for this reaction is simply K_b for the acetate ion. The salt is said to be hydrolysed, but the reaction is merely another aspect of acid–base behaviour.

Example

Calculate the pH of an aqueous solution of sodium acetate containing 1.00×10^{-2} mol kg^{-1} of CH_3COONa. (pK_b for $CH_3COO^- = 9.24$).

$$K_b = 10^{-9.24} = 5.72 \times 10^{-10}$$

$$\therefore \frac{a_{OH^-} \times a_{CH_3COOH}}{1.00 \times 10^{-2}} = \frac{(a_{OH^-})^2}{1.00 \times 10^{-2}} = 5.72 \times 10^{-10}$$

$$\therefore a_{OH^-} = (5.72 \times 10^{-12})^{\frac{1}{2}}$$

$$= 2.40 \times 10^{-6}.$$

$$\therefore a_{H_3O^+} \text{ in the solution} = \frac{1.00 \times 10^{-14}}{2.40 \times 10^{-6}}$$

$$= 4.16 \times 10^{-9}$$

and $\quad\quad\quad\quad\quad\quad pH = -\log_{10}(a_{H_3O^+})$

$$= -(\overline{9}.62)$$

$$= 8.38.$$

Unlike the aquated Na^+ ion, the aquated Fe^{3+} ion has quite a strong tendency to donate a proton to a water molecule:

$$Fe(H_2O)_6^{3+} + H_2O \rightleftharpoons Fe(H_2O)_5OH^{2+} + H_3O^+$$

The species is quite a strong acid ($pK_a = +2.6$) Iron(III) salts such as $FeCl_3$ hydrolyse strongly in water and give acidic solutions.

Buffer action

Consider a weak acid HX which reacts with water to produce X^- ions in solution:

$$HX + H_2O = X^- + H_3O^+.$$

$$K_a = \frac{a_{X^-} \times a_{H_3O^+}}{a_{HX}}$$

Rearranging:

$$K_a = a_{H_3O^+} \times \frac{a_{X^-}}{a_{HX}}$$

Hence:

$$-\log_{10} K_a = -\log_{10} a_{H_3O^+} - \log_{10} \frac{a_{X^-}}{a_{HX}}$$

or

$$pK_a = pH + \log_{10} \frac{a_{HX}}{a_{X^-}}$$

In general:

$$pH = pK_a - \log_{10} \frac{a_{acid}}{a_{conjugate\ base}} \qquad (80)$$

The ratio $a_{acid}/a_{conjugate\ base}$ is called the <u>buffer ratio</u> for the solution. $\longleftarrow$

Example

Calculate the pH of a solution of an aqueous solution containing 1.00×10^{-2} mol kg^{-1} of CH_3COOH ($pK_a = 4.76$) and 1.00×10^{-3} mol kg^{-1} of CH_3COONa.

Since CH_3COONa is completely ionised,

$$a_{CH_3COO^-} = 1.00 \times 10^{-3}.$$

Thus
$$pH = pK_a - \log_{10} \frac{a_{CH_3COOH}}{a_{CH_3COO^-}}$$

$$= 4.76 - \log_{10} \frac{1.00 \times 10^{-2}}{1.00 \times 10^{-3}}$$

$$= 3.76.$$

Comparison with Example A on p. 139 shows that the solution is made less acidic by the addition of acetate ions.

A solution containing equimolar quantities of CH_3COOH and CH_3COO^- has a pH of 4.76, since

$$pH = pK_a - \log_{10} \frac{a_{acid}}{a_{conjugate\ base}}$$

$$= 4.76 - \log_{10} 1$$

$$= 4.76.$$

The addition of small quantities of an acid (i.e. of H_3O^+ ions) or a base (OH^- ions) alters the buffer ratio slightly by the reactions

$$CH_3COO^- + H_3O^+ \rightleftharpoons CH_3COOH + H_2O,$$
or $\quad CH_3COOH + OH^- \rightleftharpoons CH_3COO^- + H_2O,$

but the corresponding change in the logarithm of the buffer ratio is so small that the pH is hardly affected. The solution is said to be buffered against change in acidity.

Example

A solution containing equimolar quantities of CH_3COOH and CH_3COONa has a pH of 4.76. A quantity of sodium hydroxide is added sufficient to convert one-half of the acetic acid to acetate ions:

$$CH_3COOH + OH^- = CH_3COO^- + H_2O.$$

Calculate the change in pH.

The effect of the OH^- ions is to make

$$a_{CH_3COO^-} = 3 \times a_{CH_3COOH}.$$

$$\therefore pH = 4.76 - \log_{10} \frac{1}{3}$$

$$= 4.76 - (\bar{1}.52)$$

$$= 5.24.$$

$$\therefore Change\ in\ pH = 0.48.$$

Buffer solutions are normally made up from approximately equimolar quantities of a weak acid and its conjugate base (for pH < 7) or a weak base and its conjugate acid (for pH > 7). The buffer ratio is clearly altered least by the addition of H_3O^+ or OH^- when its value is originally unity.

Acid-base indicators: indicator constant

A conjugate acid–base pair in which the acid and base have different colours can be used as an acid–base indicator. For example, phenolphthalein is a colourless weak acid but its anionic conjugate base is pink. The equilibrium with water in aqueous solution can be represented by:

$$HIn \quad + H_2O \rightleftharpoons \quad In^- + H_3O^+$$
$$\text{colourless} \qquad\qquad \text{pink}$$

The indicator constant, K_{HIn}, is given by

$$K_{HIn} = \frac{a_{In^-} \times a_{H_3O^+}}{a_{HIn}} \qquad (81)$$

and $\qquad pK_{HIn} = -\log_{10} K_{HIn}$ is given by

$$pK_{HIn} = pH + \log_{10} \frac{a_{HIn}}{a_{In^-}} \qquad (82)$$

For phenolphthalein $pK_{HIn} = 9.7$. (Table XVII)

In a solution of pH 9.7, therefore:

$$\log_{10} \frac{a_{HIn}}{a_{In^-}} = 0 = \log_{10} 1.0$$

and $a_{HIn} = a_{In^-}$. The solution is pink. But at pH 8.2

$$\log_{10} \frac{a_{HIn}}{a_{In^-}} = pK_{HIn} - pH = 1.5$$

and $\qquad a_{In^-} = 10^{-1.5} \times a_{HIn} = 0.032\, a_{HIn}.$

The concentration of the coloured conjugate base is so low that the colour cannot be detected. The range of pH over which phenolphthalein is an effective indicator is from pH 8.2 to pH 10.0, spanning the pK_{HIn} value. The upper limit is the point above which further deepening of the pink coloration cannot be detected by eye.

In the titration of a weak acid such as acetic acid with a strong base such as sodium hydroxide, the pH has reached a

TABLE XVII
Acid-Base Indicators

	pK_{HIn}	Transition interval	Colour	
			Acid	Base
Methyl Orange	3.4	3.2-4.4	Red	Yellow
Methyl Red	5.0	4.2-6.2	Red	Yellow
Bromothymol Blue	7.3	6.0-7.6	Yellow	Blue
Phenolphthalein	9.7	8.2-10.0	Colourless	Red

value well above 7.0 when all the acetic acid is neutralised, because the CH_3COO^- ion has itself an alkaline reaction (p. 140). For such titrations phenolphthalein is a suitable indicator. On the other hand, for the titration of a weak base against a strong acid an indicator such as methyl orange ($pK_{HIn} = 3.4$) is needed.

Examples X

1. The solubility of silver iodide in water at 298 K is 3.05 $\times 10^{-9}$. Calculate (a) the molality of the solution, ($M_{AgI} = 0.235$ kg mol^{-1}), (b) the relative activities of the Ag^+ and I^- ions in the saturated solution, (c) the solubility product at that temperature.

2. The solubility of PbS in water at 298 K is 1.70×10^{-15}. Calculate the solubility product for PbS at that temperature. $M_{PbS} = 0.239$ kg mol^{-1}.

3. For $Mn(OH)_2$ at 298 K, $K_S = 4.0 \times 10^{-14}$. Calculate the solubility of the compound in water at that temperature. M for $Mn(OH)_2 = 0.089$ kg mol^{-1}.

4. For CuCNS at 298 K, $K_S = 4.0 \times 10^{-14}$. Calculate the solubility of the compound in water at that temperature. M for CuCNS $= 0.122$ kg mol^{-1}.

5. For CuI, $K_S = 2.60 \times 10^{-12}$. Calculate the solubility of CuI ($M = 0.191$ kg mol^{-1}) in water containing 1.00×10^{-4} mol kg^{-1} of I^- ions.

6. For $PbCrO_4$, $K_S = 1.80 \times 10^{-14}$. Calculate the solubility of $PbCrO_4$ ($M = 0.323$ kg mol^{-1}) in an aqueous solution containing 5.0×10^{-4} mol kg^{-1} of Pb^{2+} ion.

7. Write the formulae of the conjugate bases of the following acids: (a) H_2SO_3 (b) H_3PO_3 (c) PH_4^+ (d) C_6H_5OH (e) $C_2H_5OH_2^+$ (f) $Cr(H_2O)_6^{3+}$.

8. Write the formulae of the conjugate acids of the following bases: (a) $C_3H_7NH_2$ (b) NO_3^- (c) CrO_4^{2-} (d) $(C_2H_5)_2O$ (e) $Al(OH)_3$.

9. Complete the following equations for protolysis reactions:

(a) $NH_3 + CH_3COOH$ $= NH_4^+ +$
(b) $CH_3COOH + CH_3COOH = CH_3COO^- +$
(c) $HSO_4^- + OH^-$ $= H_2O +$
(d) $H^- + H_2O$ $= OH^- +$

10. For formic acid, HCOOH, $pK_a = 3.79$. Use Eqn. (73) to calculate the degree of dissociation in an aqueous solution containing 1.00×10^{-2} mol kg^{-1} HCOOH.

11. Calculate pK_a (a) for an acid HA for which $K_a = 1.10 \times 10^{-5}$ (b) for an acid HX for which $K_a = 2.70 \times 10^2$.

12. Use Table XVI to place the following bases in order of increasing base strength. HSO_4^-, NH_3, $H_2PO_4^-$, CCl_3COO^-, ClO_4^-, $C_2H_5NH_2$.

13. Calculate the pH of a solution containing equimolar concentrations of $H_2PO_4^-$ and HPO_4^{2-} ions. (Use Table XVI).

14. Calculate the pH of a solution of aniline (see Table XVI) containing 1.00×10^{-3} mol kg^{-1} of $C_6H_5NH_2$ at 298 K.

15. Calculate the pH of a solution of iron(III) chloride containing 5.00×10^{-4} mol kg^{-1} of Fe^{3+}. pK_a for $Fe(H_2O)_6^{3+}$ $= +2.60$.

16. Calculate the pH of a solution of potassium cyanide containing 2.0×10^{-4} mol kg^{-1} of KCN. (Use Table XVI).

17. Calculate the weight of CH_3COONa ($M = 0.072$ kg mol^{-1}) which must be added to 1.00 kg of a solution of CH_3COOH containing 1.00×10^{-3} mol, to make a solution buffered at pH 6.00. pK_a for acetic acid $= 4.76$.

18. Calculate the weight of solid CH_3COONa which must be added to 1.00 kg of a solution containing 1.00×10^{-3} mol CH_3COOH so that a solution buffered at pH 4.00 is produced.

19. For methyl red, $pK_{HIn} = 5.0$. The range over which the indicator can be used is from pH 4.2 to pH 6.2. Calculate the ratio a_{HIn}/a_{In^-} for (a) the lower limit (b) the upper limit of the range.

Chapter 11

Oxidation and reduction

General ideas

We shall approach the quantitative aspects of oxidation and reduction by considering the well-known reaction between zinc and aqueous copper sulphate, which can be represented by the ionic equation:

$$Zn + Cu^{2+} = Zn^{2+} + Cu.$$

In this reaction zinc is oxidised by copper ions, and copper ions are reduced by zinc, two electrons being transferred from a zinc atom to a Cu^{2+} ion.

Loss of electrons = oxidation,
Gain of electrons = reduction.

Furthermore the zinc is referred to as a reducing agent and the Cu^{2+} ion as an oxidising agent in the reaction. The process involves both oxidation and reduction and is classified as a redox process.

Our concern in this chapter is to describe methods by which the effectiveness of different oxidising and reducing agents can be compared.

Reversible chemical cells

The Daniell cell once used commercially as a source of electric current has a copper plate in contact with aqueous $CuSO_4$ and a zinc plate in contact with $ZnSO_4$ as its two electrodes. It can be represented by the diagram:

$$Zn \mid Zn^{2+} \, aq \parallel Cu^{2+} \, aq \mid Cu$$

The solutions are separated by a porous partition. When the electrodes are connected by a wire a current flows in it – electrons move from the zinc to the copper. As current is withdrawn, the zinc dissolves and the $ZnSO_4$ solution becomes more concentrated; at the same time the copper plate receives a coating of copper and the $CuSO_4$ solution becomes less concentrated. However, if a storage battery connected in the external circuit forces electrons to flow in the opposition direction the chemical process is reversed; copper dissolves and zinc is plated out.

The Daniell cell is evidently a reversible chemical cell made up of two reversible electrodes or half-elements. At their surfaces there are equilibria:

$$Cu^{2+} + 2e \rightleftharpoons Cu,$$
$$Zn^{2+} + 2e \rightleftharpoons Zn.$$

Electrons flow from the zinc to the copper when the two electrodes are connected because Cu^{2+} ions are more readily discharged by electrons, i.e. they are a better oxidising agent, than Zn^{2+} ions. Thus the spontaneous cell reaction is

$$Zn + Cu^{2+} = Cu + Zn^{2+} \quad (cf.\ p.\ 148).$$

The reaction differs in only one respect from that which occurs when zinc is placed in aqueous $CuSO_4$ – its rate can be controlled by the rate at which the electrons are allowed to pass through the external circuit.

Reversible electrodes

The Cu^{2+}/Cu and Zn^{2+}/Zn electrodes are examples of cation electrodes in which a metal (the reduced state) is in equilibrium with cations formed from it (the oxidised state):

$$Metal \rightleftharpoons Cation + electron(s).$$

The most usual form of anion electrode has a metal coated with one of its insoluble salts surrounded by an electrolyte containing the anion of that salt. Silver coated with AgCl, for example, releases Cl^- ions when electrons are supplied to it:

$$AgCl\ (s) + e = Ag\ (s) + Cl^-\ (aq),$$

but Ag removes Cl^- ions from solution when electrons are withdrawn:

$$Ag\ (s) + Cl^-\ (aq) = AgCl\ (s) + e.$$

Thus the electrode is reversible to Cl^- ions.

All reversible electrodes are <u>oxidation-reduction</u> electrodes, but the term is used also in a more limited sense. An inert metal like platinum, surrounded by an aqueous solution containing some element in two different oxidation states, e.g. Fe^{2+} and Fe^{3+}, can also behave as a reversible electrode. At the platinum surface:

$$Fe^{2+} \longrightarrow Fe^{3+} \text{ when electrons are withdrawn,}$$
and $\quad Fe^{3+} \longrightarrow Fe^{2+} \text{ when electrons are supplied.}$

The oxidised and reduced species in equilibrium at a reversible electrode, e.g. Cu^{2+}/Cu or Fe^{3+}/Fe^{2+}, are called <u>redox couples</u>.

The *emf* of a cell

The <u>electromotive force</u> (*emf*) of a cell represented by a cell diagram such as

$$Zn \mid Zn^{2+} \parallel Cu^{2+} \mid Cu$$

is defined as the electric potential of the <u>right-hand</u> electrode minus the electric potential of the <u>left-hand</u> electrode under the condition that no current is flowing in the external circuit. The *emf* is measured with a <u>potentiometer</u>, which works on the principle that the *emf* produced by the cell is opposed by an exactly equal, measurable difference of potential in the external circuit. For the example above the *emf* is positive unless the ratio $a_{Cu^{2+}}/a_{Zn^{2+}}$ is extremely small (p. 155)

If the cell diagram were written

$$Cu \mid Cu^{2+} \parallel Zn^{2+} \mid Zn$$

the *emf* would be negative. The convention enables us to indicate the direction, as well as the magnitude, of the potential difference.

The double line down in the middle of the cell diagram implies that there is no difference of potential between the two electrolytes. There are practical methods for ensuring this condition but we shall not discuss the principles in this book.

The standard hydrogen electrode

The fact that the Cu^{2+}/Cu couple is positive and the Zn^{2+}/Zn couple negative in the Daniell cell indicates that Cu^{2+} is a better oxidising agent than Zn^{2+}. It is useful to have some redox couple to use as a reference standard for comparing the strengths of any oxidising and reducing agents in aqueous solution; we use the <u>standard hydrogen electrode</u> in which pure hydrogen at 101 325 N m^{-2} is passed over platinised platinum which is immersed in a solution containing H^+ (aq) ions at unit relative activity (p. 130). A suitable electrolyte is an aqueous solution of HCl with a concentration of 1.18 kmol m^{-3}. The electrode is a cation electrode reversible to H^+ ions:

$$H_2 \rightleftharpoons 2\ H^+ \text{(aq)} + 2e$$

Standard redox potential

The *emf* of the cell

$$\text{Pt, } H_2 \ (a = 1) \ | \ H^+ \text{ (aq) } (a = 1) \ \| \ Cu^{2+} \text{ (aq) } | \ Cu$$

in which the left-hand electrode is the standard hydrogen electrode, is called the <u>redox potential</u> of the Cu^{2+}/Cu couple, written E, Cu^{2+}/Cu. For the case where $a_{Cu^{2+}} = 1$, the *emf* is the <u>standard redox potential</u>, written $E°$, Cu^{2+}/Cu. Notice that the oxidised form <u>precedes</u> the oblique stroke and the reduced form <u>follows</u> it. The value of $E°$, Cu^{2+}/Cu is +0.34 V.

For the cell

$$\text{Pt, } H_2 \ | \ H^+ \text{ (aq) } (a = 1) \ \| \ Zn^{2+} \ (a = 1) \ | \ Zn$$

the *emf* is -0.76 V (i.e. the Zn^{2+}/Zn couple forms the negative electrode), implying that the spontaneous cell reaction is

$$2\ H^+ + Zn \longrightarrow H_2 + Zn^{2+}.$$

Thus $E°$, $Zn^{2+}/Zn = -0.76$ V.

For a redox couple such as Fe^{3+}/Fe^{2+}, the value of $E°$, Fe^{3+}/Fe^{2+}(+0.77 V) is defined as the *emf* of the cell

$$\text{Pt, } H_2 \ (a = 1) \ | \ H^+ \ (a = 1) \ \| \ Fe^{2+} \ (a = 1), \ Fe^{3+} \ (a = 1) \ | \ Pt.$$

Table XVIII is a list of some standard redox potentials. The reader will notice that, where the reducing agent is a metal, its order in the table is exactly the same as in the so-called electrochemical series of metals.

Standard *emfs*

The standard *emf* of a cell, *emf°*, is obtained by subtracting $E°$ for the left-hand electrode from $E°$ for the right-hand one. Thus for the cell

$$\text{Zn} \ | \ Zn^{2+} \ (a = 1) \ \| \ Ag^+ \ (a = 1) \ | \ Ag$$

TABLE XVIII
Some Standard Redox Potentials at 298 K (pH = 0)

Electrode reaction		Redox couple	$E°/V$
$F_2 + 2e$	$= 2 F^-$	F_2/F^-	+2.87
$MnO_4^- + 8 H^+ + 5e$	$= Mn^{2+} + 4 H_2O$	MnO_4^-/Mn^{2+}	+1.51
$Cl_2 + 2e$	$= 2 Cl^-$	Cl_2/Cl^-	+1.36
$Cr_2O_7^{2-} + 14 H^+ + 6e$	$= 2 Cr^{3+} + 7 H_2O$	$Cr_2O_7^{2-}/Cr^{3+}$	+1.33
$MnO_2 + 4 H^+ + 2e$	$= Mn^{2+} + 2 H_2O$	MnO_2/Mn^{2+}	+1.23
$Br_2 + 2e$	$= 2 Br^-$	Br_2/Br^-	+1.07
$NO_3^- + 3 H^+ + 2e$	$= HNO_2 + H_2O$	NO_3^-/HNO_2	+0.94
$Ag^+ + e$	$= Ag$	Ag^+/Ag	+0.80
$Fe^{3+} + e$	$= Fe^{2+}$	Fe^{3+}/Fe^{2+}	+0.77
$I_2 + 2e$	$= 2 I^-$	I_2/I^-	+0.54
$Cu^{2+} + 2e$	$= Cu$	Cu^{2+}/Cu	+0.34
$Sn^{4+} + 2e$	$= Sn^{2+}$	Sn^{4+}/Sn^{2+}	+0.15
$2 H^+ + 2e$	$= H_2$	H^+/H_2	0.00
$Sn^{2+} + 2e$	$= Sn$	Sn^{2+}/Sn	−0.14
$Ni^{2+} + 2e$	$= Ni$	Ni^{2+}/Ni	−0.25
$Fe^{2+} + 2e$	$= Fe$	Fe^{2+}/Fe	−0.44
$Zn^{2+} + 2e$	$= Zn$	Zn^{2+}/Zn	−0.76
$Mg^{2+} + 2e$	$= Mg$	Mg^{2+}/Mg	−2.37
$Na^+ + e$	$= Na$	Na^+/Na	−2.71
$Ca^{2+} + 2e$	$= Ca$	Ca^{2+}/Ca	−2.87
$Li^+ + e$	$= Li$	Li^+/Li	−3.04

$$emf° = E°, Ag^+/Ag - E°, Zn^{2+}/Zn$$

$$= +0.80 \ V - (-0.76 \ V)$$

$$= +1.56 \ V$$

Example

Calculate the standard *emf* of the cell

$$Pt \ \left| \ \begin{array}{c} Sn^{4+} \ (a = 1) \\ Sn^{2+} \ (a = 1) \end{array} \ \right\| \ \begin{array}{c} Fe^{3+} \ (a = 1) \\ Fe^{2+} \ (a = 1) \end{array} \ \right| \ Pt$$

Write down the spontaneous cell reaction.

From Table XVIII, $E°$, $Sn^{4+}/Sn^{2+} = +0.15$ V and $E°$, $Fe^{3+}/Fe^{2+} = +0.77$ V.

$$\therefore \ emf° = E°, \ Fe^{3+}/Fe^{2+} - E°, \ Sn^{4+}/Sn^{2+}$$
$$= +0.77 \text{ V} - 0.15 \text{ V}$$
$$= +0.62 \text{ V}.$$

Since the Fe^{3+}/Fe^{2+} is the positive (electron-accepting) electrode the spontaneous electrode processes are:

$$Fe^{3+} + e = Fe^{2+} \qquad \text{(a)}$$
and $$Sn^{2+} = Sn^{4+} + 2e. \qquad \text{(b)}$$

The cell reaction is obtained by adding (b) to 2(a) — there must be no unbalanced electron in the full equation:

$$2 \ Fe^{3+} + Sn^{2+} = Sn^{4+} + 2 \ Fe^{2+}.$$

Comparison of oxidising power

In Table XVIII the redox couples are listed in order of decreasing $E°$ values. The strongest oxidising agent in the list is fluorine; the species down the left of the table are progressively weaker oxidising agents as the $E°$ values decrease, and the weakest is the Li^+ ion. The strongest reducing agent shown in the table is lithium; the species up the right of the table are progressively weaker reducing agents as the $E°$ values increase, the weakest is the F^- ion.

The fact that $E°$, Cl_2/Cl^- is higher than $E°$, Br_2/Br^-, for example, indicates that chlorine is a stronger oxidising agent than bromine. Thus chlorine will oxidise a bromide to bromine:

$$Cl_2 + 2 \ Br^- = Br_2 + 2 \ Cl^-.$$

Similarly Ag^+ will oxidize Fe^{2+} to Fe^{3+} and be itself reduced to Ag because $E°$, $Ag^+/Ag > E°$, Fe^{3+}/Fe^{2+}.

Example

Use Table XVIII to place the metals iron, magnesium, calcium, silver and tin in order of reducing power. Examination of the table shows that

$$E°, \ Ag^+/Ag > E°, \ Sn^{2+}/Sn > E°, \ Fe^{2+}/Fe > E°,$$
$$Mg^{2+}/Mg > E°, \ Ca^{2+}/Ca.$$

Thus the order of reducing power of the metals is

$$Ca > Mg > Fe > Sn > Ag.$$

Standard *emf* and $\triangle G°$

A chemical cell converts chemical energy into electrical energy, and, if it operates under reversible conditions at constant temperature and pressure, the maximum electrical work is done. Consider a reversible chemical cell of standard electromotive force *emf°* in which z moles of electrons are transferred in one mole of the cell reaction. The reversible (i.e. maximum) work done in the external circuit is

$$w \; = \; zF \; (emf°).$$

Thus the change in G, the Gibbs function of the chemical system (p. 88), is given by

$$\Delta G° \; = \; -zF \; emf° \tag{83}$$

For the reaction

$$2 \; Fe^{3+} + Sn^{2+} \; = \; 2 \; Fe^{2+} + Sn^{4+}$$

$emf° = +0.62$ V (p. 150) and $z = 2$. Since F, Faraday's constant, is 9.65×10^4 C mol^{-1},

$$\Delta G° \; = \; -2 \times 0.62 \text{ V} \times 9.65 \times 10^4 \text{ C mol}^{-1}$$
$$= \; -1.19 \times 10^5 \text{ J mol}^{-1}$$
$$= \; -119 \text{ kJ mol}^{-1}$$

Example

For the cell

$$Ni \; | \; Ni^{2+} \; \| \; Cu^{2+} \; | \; Cu$$

at 298 K, $emf = +0.59$ V. Calculate $\Delta G°$ for the reaction

$$Ni + Cu^{2+} \; = \; Ni^{2+} + Cu.$$

The sign of the *emf* shows that copper is the positive electrode, i.e. it accepts electrons supplied by the nickel electrode:

$$2e + Cu^{2+} \; = \; Cu$$
$$Ni \; = \; Ni^{2+} + 2e$$

Thus, the spontaneous cell reaction is:

$$Ni + Cu^{2+} \; = \; Ni^{2+} + Cu$$

and

$$z \; = \; 2$$

$$\therefore \; \Delta G° \; = \; -zF \; (emf°)$$

$$= -2 \times 9.65 \times 10^4 \text{ C mol}^{-1} \times 0.59 \text{ V}$$

$$= -114 \text{ kJ mol}^{-1}$$

for the reaction under the specified conditions.

Equilibrium constants of redox reactions

Consider the cell

$$\text{Pt} \mid \text{Fe}^{3+} (a = 1), \text{ Fe}^{2+} (a = 1) \parallel \text{Ag}^+ (a = 1) \mid \text{Ag}.$$

The cell reaction is

$$\text{Ag}^+ + \text{Fe}^{2+} = \text{Ag} + \text{Fe}^{3+},$$

$z = 1$ for the reaction as written and $emf^\circ = +0.80 \text{ V} - 0.77 \text{ V}$ $= +0.03 \text{ V}$ at 298 K (Table XVIII).

If a current is taken from the cell for some time, silver is plated out, the activity of Ag^+ falls and the ratio $a_{\text{Fe}^{3+}}/a_{\text{Fe}^{2+}}$ increases. The consequence is that the emf of the cell falls below the standard value, emf°, eventually reaching zero. At this point no further current can be drawn from the cell; the reactants are at their equilibrium activities.

The relation between ΔG for a reaction and the activities of the reactants and products is given by the reaction isotherm:

i.e. $$\Delta G = \Delta G^\circ + RT \ln \frac{a_{\text{Ag}} \times a_{\text{Fe}^{3+}}}{a_{\text{Ag}^+} \times a_{\text{Fe}^{2+}}} \quad \text{(p. 90)}$$

for the above reaction.

Thus K, the equilibrium constant, which is the value of $\frac{a_{\text{Ag}} \times a_{\text{Fe}^{3+}}}{a_{\text{Ag}^+} \times a_{\text{Fe}^{2+}}}$ when equilibrium is reached (i.e. when $\Delta G = 0$) is given by:

$$-RT \ln K = -zF (emf^\circ)$$

$$\therefore \ln K = \frac{zF (emf^\circ)}{RT} \quad (84)$$

In the example given above

$$\log_{10} K = \frac{1 \times 9.65 \times 10^4 \text{ C mol}^{-1} \times 0.03 \text{ V}}{8.314 \text{ J K}^{-1} \text{ mol}^{-1} \times 298 \text{ K} \times 2.303}$$

$$= 0.51$$

$$\therefore K = 10^{0.51} = 3.2 \text{ at } 298 \text{ K}.$$

Example

Use Table XVIII to calculate the equilibrium constant of the reaction

$$2 \ Fe^{3+} + 2 \ I^- \ = \ 2 \ Fe^{2+} + I_2 \text{ at 298 K.}$$

For the reaction as written, $z = 2$. Since $E°$, $Fe^{3+}/Fe^{2+} = +0.77$ V and $E°$, $I_2/I^- = +0.54$ V, *emf*° for the cell in which the reaction occurs reversibly is $+0.77$ V $- 0.54$ V $= +0.23$ V. Thus K for the reaction is given by

$$\log_{10} K \ = \ \frac{2 \times 9.65 \times 10^4 \ C \ mol^{-1} \times 0.23 \ V}{8.314 \ J \ K^{-1} \ mol^{-1} \times 298 \ K \times 2.303}$$

$$= \ 7.78$$

$$\therefore K \ = \ 10^{7.78}$$

$$= \ 6.0 \times 10^7.$$

Note that the equilibrium constant for a redox reaction is large unless the difference between the $E°$ values for the two redox couples involved is very small, as in the case of Fe^{3+}/Fe^{2+} and Ag^+/Ag.

The dependence of electrode potentials on ionic activities

Consider the cell:

$$Pt, \ H_2 \ (a = 1) \ | \ H^+ \ (a = 1) \ \| \ Fe^{3+}, \ Fe^{2+} \ | \ Pt$$

in which the activities of the Fe^{3+} and Fe^{2+} ions are not equal. Suppose the *emf* of the cell to be E, then the value of ΔG for the cell reaction:

$$Fe^{3+} + \tfrac{1}{2}H_2 \ = \ Fe^{2+} + H^+$$

is given by:

$$\Delta G \ = \ -zFE \quad \text{(p. 152).}$$

From the reaction isotherm (p. 90)

$$\Delta G \ = \ \Delta G° + RT \ \ln \frac{a_{Fe^{2+}} \times \ a_{H^+}}{a_{Fe^{3+}} \times \ (a_{H_2})^{\frac{1}{2}}}$$

Thus E for the above cell is given by

$$E \ = \ E° - \frac{RT}{zF} \ \ln \frac{a_{Fe^{2+}}}{a_{Fe^{3+}}}$$

$$= E^\circ + \frac{RT}{zF} \ln \frac{a_{Fe^{3+}}}{a_{Fe^{2+}}}$$

This is the relation between the redox potential (p. 149) and the standard redox potential.

For the general case:

$$E = E^\circ + \frac{RT}{zF} \ln \frac{a_{OXID.}}{a_{RED.}} \qquad (85)$$

where $a_{OXID.}$ is the activity of the oxidised form, $a_{RED.}$ is the activity of the reduced form of the couple, and z is the numerical difference between the oxidation state (p. 182) of the oxidised form of the element and that of the reduced form.

Example

Calculate the electrode potential E, Cu^{2+}/Cu at 298 K for a copper plate immersed in a solution of copper sulphate in which $a_{Cu^{2+}} = 1.00 \times 10^{-3}$.

$$E = E^\circ + \frac{RT}{zF} \ln \frac{a_{Cu^{2+}}}{a_{Cu}}$$

Since the copper is a pure metal in its standard state, $a_{Cu} = 1.0$.

$$\therefore E = +0.34 \text{ V} + \left(\frac{8.314 \text{ J K}^{-1} \text{ mol}^{-1} \times 298 \text{ K}}{2 \times 9.65 \times 10^4 \text{ C mol}^{-1}} \times 2.303 \log_{10} (10^{-3}) \right)$$

$$= +0.34 \text{ V} - 0.09 \text{ V}$$

$$= +0.25 \text{ V}$$

Concentration cells

It follows from Eqn. (85) that in a cell such as:

$$Cu \mid Cu^{2+}, (a = a_1) \parallel Cu^{2+}, (a = a_2) \mid Cu$$

the *emf* is given by

$$emf = \frac{RT}{zF} \ln \frac{a_2}{a_1}$$

Such a cell is a concentration cell. If $a_2 > a_1$, the *emf* is positive, and electrons flow from left to right in the external circuit; copper is plated at the right and a_2 decreases, while copper dissolves at the left and a_1 increases.

Examples XI

1. Use Table XVIII to calculate the standard *emfs* of the following cells at 298 K.

(a) $Ag \mid Ag^+ \parallel H^+, MnO_4^-, Mn^{2+} \mid Pt$
(b) $Sn \mid Sn^{2+} \parallel Sn^{2+}, Sn^{4+} \qquad \mid Pt$
(c) $Pt \mid I_2, I^- \parallel NO_3^-, H^+, HNO_2 \mid Pt$
(d) $Ni \mid Ni^{2+} \parallel Br^-, Br_2 \qquad\qquad \mid Pt$

2. Which of the following species can be oxidised by $Cr_2O_7^{2-}$ in aqueous solution at pH 0?

$$Cl^-, \ I^-, \ Ag, \ Mn^{2+}.$$

3. Which of the following species in aqueous solution at pH 0 can be reduced by metallic silver?

$$Br_2, \ Cu^{2+}, \ Sn^{4+}, \ H^+$$

4. Write equations for the cell reactions which occur spontaneously in the cells.

(a) $Pt \mid Fe^{3+}, Fe^{2+} \qquad\quad \parallel H^+, MnO_4^-, Mn^{2+} \mid Pt$
(b) $Pt \mid I_2, I^- \qquad\qquad\quad\ \parallel Fe^{3+}, Fe^{2+} \qquad\quad \mid Pt$
(c) $Pt \mid NO_3^-, H^+, HNO_2 \parallel H^+, Mn^{2+} \qquad\quad \mid MnO_2$
(d) $Zn \mid Zn^{2+} \qquad\qquad\quad \parallel Ni^{2+} \qquad\qquad\quad\ \mid Ni$

5. Calculate the value of $\Delta G°$ at 298 K for each of the reactions in Q.4 above.

6. Use Table XVIII to calculate the equilibrium constants of the following reactions at 298 K.

(a) $Sn^{4+} + Sn \qquad\qquad\qquad\quad = 2\ Sn^{2+}$
(b) $3\ Cl_2 + 2\ Cr^{3+} + 7\ H_2O = 6\ Cl^- + Cr_2O_7^{2-} + 14\ H^+$
(c) $MnO_2 + 2\ Br^- + 4\ H^+ \quad = Mn^{2+} + 2\ H_2O + Br_2.$

7. Calculate the redox potentials of the following redox couples at 298 K.

(a) Zn^{2+} ($a = 1.0 \times 10^{-2}$) / Zn
(b) $\frac{1}{2}\ Cl_2$ ($a = 1$) / Cl^- ($a = 0.5$)
(c) Sn^{4+} ($a = 0.10$) / Sn^{2+} ($a = 0.010$).

8. Calculate the *emfs* of the following concentration cells at 298 K.

(a) $Zn \mid Zn^{2+}$ $(a = 0.10) \parallel Zn^{2+}$ $(a = 0.010) \mid Zn$

(b) Pt, Cl_2 $(a = 1.0) \mid Cl^-$ $(a = 1.0)$ $\parallel Cl^-$ $(a = 1.0 \times 10^{-3}) \mid$
Cl_2 $(a = 1.0)$, Pt

(c) $Pt \mid Sn^{4+}$ $(a = 1.0)$, Sn^{2+} $(a = 1.0) \parallel Sn^{4+}$ $(a = 1.0)$, Sn^{2+}
$(a = 1.0 \times 10^{-2}) \mid Pt$

9. The *emf* of the following concentration cell at 298 K is +0.125 V. Calculate a_2.

$$Cu \mid Cu^{2+} \ (a = 1.0 \times 10^{-4}) \parallel Cu^{2+} \ (a = a_2) \mid Cu.$$

10. The *emf* of the following cell is -0.52 V at 298 K. Calculate a_{H^+} and hence the pH of the right-hand electrolyte.

$$Pt, H_2 \ (a = 1.0) \mid H^+ \ (a = 1.0) \parallel H^+ \mid H_2 \ (a = 1.0), \ Pt.$$

Chapter 12

Electrolytic conduction

The nature of conductance in an electrolyte

If two pieces of copper foil are immersed in aqueous $CuSO_4$ solution and connected to the poles of a battery, the cathode becomes coated with copper but the anode erodes away. The experiment distinguishes <u>electrolytic conduction</u> in which the current is carried by ions, for example the Cu^{2+} cations and the SO_4^{2-} anions in the above case, from <u>metallic conduction</u>, in metallic copper for example, in which the flow of electricity is interpreted as a stream of electrons.

Experiments to demonstrate the nature of electrolytic conduction were devised by Lodge. He made a solution of $CuSO_4$ in warm aqueous gelatine, poured it into a U–tube and allowed it to set. Above the blue jelly he poured a solution in aqueous gelatine of a colourless electrolyte like KNO_3 and finally a similar solution of a barium salt. When a potential difference of 100 V was applied to electrodes immersed in either limb, the blue colour moved bodily towards the cathode, and a white turbidity eventually appeared under the anode. Clearly the Cu^{2+} ions move towards the negative pole and the SO_4^{2-} ions towards the positive one, forming a precipitate of $BaSO_4$ as they meet the Ba^{2+} ions moving towards them.

In earlier experiments Faraday studied quantitatively the chemical changes which occurred at the electrodes, and summarised his results in two laws:

(i) The amount of chemical action produced by a current is proportional to the quantity of electricity passed.

(ii) The masses of different substances deposited or dissolved by the same quantity of electricity are proportional to their chemical equivalent weights.

In modern terms we should say that the charge carried by one mole of ions of charge $+ze$ is $+zF$, where F is known as Faraday's constant, 9.649×10^4 C mol^{-1}, which represents the ⬅ charge on one mole of electrons; thus the (negative) charge on one electron is $F/N_A = 9.649 \times 10^4$ C mol$^{-1}/6.023 \times 10^{23}$ mol^{-1} $= 1.602 \times 10^{-19}$ C.

It follows from Faraday's laws that if a current is passed through a series of electrolytic cells in series the fraction

$$\frac{\text{amount of ionic species discharged}}{\text{charge carried by ion}}$$

is the same at every electrode.

Example

An electric current of 0.34 A was passed through (a) a solution of copper sulphate between copper plates, (b) a solution of silver nitrate between silver plates, (c) a solution of potassium iodide between platinum plates, for 30.0 minutes. Calculate the weights of (i) copper deposited at the cathode in (a), (ii) silver deposited at the cathode in (b), (iii) iodine liberated at the anode in (c).

$$M\,(\text{Cu}^{2+}) = 0.063\ 55 \text{ kg mol}^{-1},$$
$$M\,(\text{Ag}^{+}) = 0.107\ 87 \text{ kg mol}^{-1},$$
$$M\,(\text{I}^{-}) = 0.126\ 90 \text{ kg mol}^{-1}.$$

Quantity of electricity passed through the solutions

$$= 30.0 \times 60 \text{ s} \times 0.34 \text{ A}$$

$$= 612 \text{ C}.$$

$\therefore$ Weight of Cu^{2+} discharged

$$= \frac{612 \text{ C}}{96.5 \times 10^3 \text{ C mol}^{-1}} \times \frac{0.063\ 55 \text{ kg mol}^{-1}}{2}$$

$$= 2.015 \times 10^{-4} \text{ kg} = 201.5 \text{ mg}.$$

Weight of Ag^{+} discharged

$$= \frac{612 \text{ C}}{96.5 \times 10^3 \text{ C mol}^{-1}} \times \frac{0.107\ 87 \text{ kg mol}^{-1}}{1}$$

$$= 6.843 \times 10^{-4} \text{ kg} = 684.3 \text{ mg}.$$

Weight of I^- discharged

$$= \frac{612 \text{ C}}{96.5 \times 10^3 \text{ C mol}^{-1}} \times \frac{0.126\ 90 \text{ kg mol}^{-1}}{1}$$

$$= 8.050 \times 10^{-4} \text{ kg} = 805.0 \text{ mg}.$$

∴ Weights of copper, silver and iodine released are respectively (i) 201.5 mg, (ii) 684.3 mg, (iii) 805.0 mg.

Ionic mobilities

The velocity, v_i, of an ion can be measured in some cases by experiments similar to those of Lodge. The electric mobility, u_i, is defined by the equation

$$u_i = \frac{v_i}{E} \qquad (86)$$

where E is the electric potential gradient in the solution.

The molar conductivity of an ion, λ_i, carrying a charge $+z_i e$ is defined by the equation:

$$\lambda_i = z_i u_i F \qquad (87)$$

The value of λ_i for a particular ion depends on the viscosity of the solvent, the temperature and the concentration (p. 161). Accurate methods of determining λ_i values depend on the measurement of the electric resistance of electrolyte solutions, followed by the evaluation of the fraction of the total electrolytic current carried by a particular kind of ion in the solution at the same concentration and temperature (p. 163).

Determining the molar conductivity of an electrolyte

A conductivity cell consists of platinum electrodes immersed parallel to one another in the electrolyte to be studied. If there is a distance l between the plates, each of cross-section a, and the resistance of the cell is R, then the resistivity of the solution, ρ, is given by

$$\rho = \frac{Ra}{l} \qquad (88)$$

The resistance R is measured with the well-known Wheatstone bridge. To prevent the formation of bubbles or of concentration gradients at the electrodes, an alternating current is used. The ratio l/a is called the cell constant and is usually determined by measuring the resistance of the cell when it is filled with a standard solution of known resistivity.

The reciprocal of ρ is the electrolytic conductivity, κ.
The molar conductivity of the electrolyte, λ, is defined by the
equation

$$\lambda = \kappa/c = \frac{1}{\rho c} \tag{89}$$

where c is the concentration of the electrolyte.

Example

In a conductivity cell the parallel electrodes are each of
area 1.2×10^{-4} m^2, and the distance between them is 3.0×10^{-2}
m. When an electrolyte of concentration 200.0 mol m^{-3} is
placed in the cell the resistance is 60.0 Ω at 290 K. Calculate
(a) the resistivity, (b) the conductivity and hence (c) the molar
conductivity of the electrolyte under the conditions of the experiment.

$$\rho = \frac{Ra}{l} = \frac{60.0 \ \Omega \times 1.2 \times 10^{-4} \ \text{m}^2}{3.0 \times 10^{-2} \ \text{m}}$$

$$= 2.4 \times 10^{-1} \ \Omega \ \text{m}.$$

$$\therefore \ \kappa = \frac{1}{\rho} = 4.17 \ \Omega^{-1} \ \text{m}^{-1}.$$

Note: There is always some contribution to the conductivity
from the H_3O^+ and OH^- ions of the water, but for carefully dis-
tilled water κ is only about $7 \times 10^{-5} \ \Omega^{-1} \ \text{m}^{-1}$. We can therefore
ignore this contribution in the present case.

Then

$$\lambda = \kappa/c$$

$$= \frac{4.17 \ \Omega^{-1} \ \text{m}^{-1}}{200.0 \ \text{mol m}^{-3}}$$

$$= 2.09 \times 10^{-2} \ \Omega^{-1} \ \text{m}^2 \ \text{mol}^{-1}.$$

The variation of λ with concentration in strong electrolytes

The value of λ for a strong electrolyte increases slightly
with dilution. The further the positive and negative ions are
apart the less they hinder one another as they move in opposite
directions in an applied electric field. The relation between λ
at a particular concentration c, and λ_0, the molar conductivity
at zero dilution, is given for a strong electrolyte by an empirical equation

$$\lambda = \lambda_0 - kc^{\frac{1}{2}} \tag{90}$$

where k is an experimental constant. The value of λ_0 can therefore be found by plotting λ/Ω^{-1} m^2 mol^{-1} against $[c/\text{mol}$ m$^{-3}]^{\frac{1}{2}}$ and extrapolating to zero concentration (see Q. 7 and 8 on p. 168 and Fig. 36 on p. 167).

Kohlrausch's law of independent mobilities

Kohlrausch (1875) observed that λ_0 for a potassium salt in aqueous solution at 291 K was always $\sim 2.15 \times 10^{-3}$ Ω^{-1} m^2 mol^{-1} greater than λ_0 for the corresponding sodium salt at the same temperature, whatever the common anion. The behaviour can be explained if λ_0 is assumed to be the sum of two independent terms, one characteristic of the cation and the other of the anion. Thus for a 1:1 electrolyte, to take the simplest example:

$$\lambda_0 = \lambda_0{}^+ + \lambda_0{}^- \tag{91}$$

where $\lambda_0{}^+$ and $\lambda_0{}^-$ are the limiting values of the molar conductivities of the respective ions at zero dilution.

Determining the fraction of total current carried by a particular ion

A sharp boundary between an aqueous electrolyte CA and another MA in a narrow vertical tube can be distinguished if the layers differ in colour or in refractive index. Provided the C^+ ion is by nature a faster-moving ion than the M^+, the boundary remains sharply defined as the cations move up towards the cathode when an electric potential difference is applied (Fig. 34). The M^+ ion does not overtake the C^+, but neither does it fall behind, because if the liquid following behind the boundary becomes more dilute its higher resistance and the resulting increase in the electric potential gradient cause the speed of M^+ to increase (p. 160).

Suppose the boundary moves a distance x in a tube of cross-section a when a quantity Q of electricity is passed through the tube. If the concentration of MA is c the quantity of electricity carried past a particular point in the tube by M^+ ions is $xacF$. Thus the fraction of the total current carried by the M^+ ion is

$$t_{M^+} = \frac{xacF}{Q} \tag{92}$$

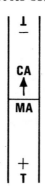

Fig. 34. Representation of
moving boundary.

The quantity t_{M^+} is called the transport number of the M^+ ion ⬅
in the solution.

Example

In a glass tube with a uniform cross-section of 1.00×10^{-5}
m^2 the boundary of a solution of HCl containing 10.0 mol m^{-3}
moved 17.0 cm towards the cathode when a current of 10.0 mA
was passed for 200.0 s. Calculate t_{H^+} under these conditions.

$$t_{H^+} = \frac{xacF}{Q}$$

$$= \frac{1.70 \times 10^{-1}\, m \times 1.00 \times 10^{-5}\, m^2 \times 10.0\, mol\, m^{-3} \times 9.649 \times 10^4\, C\, mol^{-1}}{1.0 \times 10^{-2}\, A \times 200.0\, s}$$

$$= 0.82.$$

As t_{H^+} is the fraction of the current carried by the H^+ ion, the
fraction carried by the anion, t_{Cl^-}, is

$$1 - 0.82 = 0.18.$$

Moving boundary experiments like those described above enable
the values of λ_0^+ and λ_0^- to be obtained, since

$$\lambda_0^{\ +} = t_+ \times \lambda_0 \text{ and } \lambda_0^{\ -} = t_- \times \lambda_0. \qquad (93)$$

For HCl at 298 K, $\lambda_0 = 42.6 \times 10^{-3}\, \Omega^{-1}\, m^2\, mol^{-1}$ and t_{H^+}
$= 0.822.$

$$\therefore \lambda_0^{\ +} \text{ for } H^+ \text{ is} \qquad 0.822 \times 42.6 \times 10^{-3}\, \Omega^{-1}\, m^2\, mol^{-1}$$

$$= 35.0 \times 10^{-3}\, \Omega^{-1}\, m^2\, mol^{-1}.$$

Similarly λ_0^- for Cl$^-$ $= 0.178 \times 42.6 \times 10^{-3} \; \Omega^{-1} \; m^2 \; mol^{-1}$

$\qquad\qquad\qquad\qquad = 7.6 \times 10^{-3} \; \Omega^{-1} \; m^2 \; mol^{-1}.$

Table XIX lists the molar conductivities at zero concentration of some ions (298 K). The H_3O^+ and OH^- ions appear much more mobile than the others, probably because rapid exchanges of protons with water molecules occur as well as ordinary ionic migration during the process of conduction.

Conductometric titrations

The exceptionally high molar conductivities of H^+ and OH^- enable acid-base titrations to be performed in a conductivity cell on solutions which are too highly coloured to permit the use of ordinary indicators (p. 143). A dilute solution of an alkali is placed in the cell and the resistance, R, is measured. A fairly concentrated acid is added in measured quantities – best results are obtained if the volume remains almost constant – and R is redetermined after each addition of acid. Figure 35 is a typical graph for the variation of R when a strong acid is added to a strong base.

The line AX represents the fall in conductivity caused by the replacement of fast OH^- ions in the cell by slow X^- ions

$$OH^- \quad + \quad H^+ + X^- \quad = \quad H_2O \quad + \quad X^-$$

(in cell)   (acid added)   (in cell)

The line XB represents the rise in conductivity as an excess of very fast H^+ ions is added. The value of V corresponding to the point X therefore represents the volume of acid required to neutralise the alkali.

Determining the solubility of a sparingly soluble salt

The conductivity of a solution saturated with $BaSO_4$ at 298 K was found by measurement of its resistance in a conductivity cell to be $4.20 \times 10^{-4} \; \Omega^{-1} \; m^{-1}$. The water itself had a conductivity of $1.05 \times 10^{-4} \; \Omega^{-1} \; m^{-1}$. Thus the conductivity due to the Ba^{2+} and SO_4^{2-} ions is $3.15 \times 10^{-4} \; \Omega^{-1} \; m^{-1}$. But since:

$$\kappa = \lambda c,$$

and as the solution can be considered to approximate to zero concentration,

TABLE XIX

Molar Conductivities at 298 K of Ions at Zero Concentration

	$10^3\lambda/\Omega^{-1}$ m^2 mol^{-1}		$10^3\lambda/\Omega^{-1}$ m^2 mol^{-1}
H$^+$	34.98	OH$^-$	19.85
K$^+$	7.35	Cl$^-$	7.63
Ag$^+$	6.19	Br$^-$	7.84
Na$^+$	5.01	I$^-$	7.68
Li$^+$	3.87	NO$_3^-$	7.14
Ba^{2+}	12.73	CH$_3$COO$^-$	4.09
Mg^{2+}	10.61	SO$_4^{2-}$	15.92

$$c = \frac{\kappa}{\lambda_0\ (BaSO_4)} = \frac{\kappa}{\lambda_0\ (Ba^{2+}) + \lambda_0\ (SO_4^{2-})}$$

$$= \frac{3.15 \times 10^{-4}\ \Omega^{-1}\ m^{-1}}{28.65 \times 10^{-3}\ \Omega^{-1}\ m^2\ mol^{-1}}$$

$$= 1.10 \times 10^{-2}\ mol\ m^{-3}.$$

In very dilute solution this is approximately 1.10×10^{-5} mol kg^{-1} (p. 207). Since M (BaSO$_4$) = 0.233 kg mol^{-1} the solubility of BaSO$_4$ at 298 K is

$$0.233\ kg\ mol^{-1} \times 1.10 \times 10^{-5}\ mol\ kg^{-1} = 2.56 \times 10^{-6}.$$

The conductivity of weak electrolytes

Experiments on the variation of λ with concentration for weak electrolytes (p. 167) show that the ratio $\frac{\lambda}{\lambda_0}$ is approximately proportional to $c^{-\frac{1}{2}}$. As λ_0 is a constant for a particular temperature, λ is approximately proportional to $c^{-\frac{1}{2}}$, or, expressed in another way, $\lambda c^{\frac{1}{2}}$ = a constant. Figure 36 shows a typical graphical relation between λ/Ω^{-1} m^2 mol^{-1} and $[c/\text{mol m}^{-3}]^{\frac{1}{2}}$ for a weak electrolyte (a) and a strong electrolyte (b). The curve for the weak electrolyte approximates to a rectangular hyperbola (like the 'Boyle's law' plot of V against p) whereas for the strong electrolyte there is a straight-line relation (Eqn. 90, p. 161).

Arrhenius (1887) argued that the ratio $\frac{\lambda}{\lambda_0}$ represented the degree of ionisation, α, of the weak electrolyte. Using a

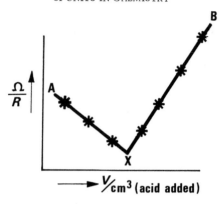

Fig. 35. Conductometric titration of strong alkali with strong acid.

derivation similar to that on p. 135, Ostwald obtained an equation for the relation between degree of ionisation and concentration

$$\frac{\alpha^2 c}{1 - \alpha} = K \qquad (94)$$

the so-called Ostwald dilution law. If we substitute $\frac{\lambda}{\lambda_0}$ for α, the equation becomes

$$\frac{\lambda^2 c}{\lambda_0 (\lambda_0 - \lambda)} = K \qquad (95)$$

where K is a constant having the units of concentration, i.e. mol m^{-3} in the SI. This equation is obeyed closely by weak electrolytes in dilute solution. The reason for the great difference in behaviour between strong and weak electrolytes in the variation of λ with c is that in strong electrolytes the increase of λ on dilution is due entirely to the increased mobility of the ions (p. 161) whereas in weak electrolytes the increase is due almost entirely to the increased dissociation of the solute into ions. In 'intermediate' electrolytes the variation of λ with c is complex, but fortunately not many electrolytes can be so classified, and we shall not deal with them here. However, with intermediate and strong electrolytes the ratio $\frac{\lambda}{\lambda_0}$ cannot be equated to the degree of dissociation α even to a first approximation, and if the ratio is used it is best called the conductivity ratio.

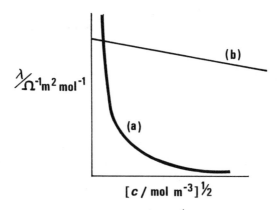

Fig. 36. Variation of λ with $c^{\frac{1}{2}}$ for (a) weak electrolyte (b) strong electrolyte.

Determination of α for a weak electrolyte

The resistance of a cell with a cell constant 13.7 m^{-1}, containing a solution of acetic acid of concentration 15.81 mol m^{-3}, was found to be 655 Ω. Use Table XIX to calculate λ_0 (CH_3COOH) and hence calculate the conductivity ratio for acetic acid at the given concentration.

From Eqn. (88), $\rho = \dfrac{655\ \Omega}{13.7\ m^{-1}}$

Thus from Eqn. (89), $\lambda = \dfrac{13.7\ m^{-1}}{655\ \Omega \times 15.81\ mol\ m^{-3}}$

$$= 1.32 \times 10^{-3}\ \Omega^{-1}\ m^2\ mol^{-1}.$$

But from Eqn. (91) and Table XIX

$\lambda_0\ (CH_3COOH) = (34.98 + 4.09) \times 10^{-3}\ \Omega^{-1}\ m^2\ mol^{-1}$

$$= 39.07 \times 10^{-3}\ \Omega^{-1}\ m^2\ mol^{-1}$$

$$\therefore \frac{\lambda}{\lambda_0} = \frac{1.32 \times 10^{-3}\ \Omega^{-1}\ m^2\ mol^{-1}}{39.07 \times 10^{-3}\ \Omega^{-1}\ m^2\ mol^{-1}} = 3.38 \times 10^{-2}.$$

For such a weak electrolyte we can substitute α, the degree of ionisation, for $\dfrac{\lambda}{\lambda_0}$. Thus α for the solution is about 3.38×10^{-2}. As a solution containing 15.81 mol m^{-3} has a molality of about 0.01581 mol kg^{-1} (p. 207), substitution in Eqn. (72) (p. 135) gives an approximate value for the dissociation constant for acetic acid:

$$K = \frac{(3.38 \times 10^{-2})^2 \times 0.01581 \text{ mol kg}^{-1}}{(1 - 0.0338) \times 1 \text{ mol kg}^{-1}}$$

$$= 1.87 \times 10^{-5}.$$

This value should be compared with that obtained (p. 135) after correction for the relative activities (p. 131) of the ions.

Examples XII

1. For I_2, $M = 0.2538$ kg mol^{-1}. Calculate the mass of iodine released in the electrolysis of aqueous KI with platinum electrodes when 50.0 C of electricity is passed through the electrolyte.

2. A water voltameter and a copper voltameter were connected in series. Calculate the mass of O_2 ($M = 0.03200$ kg mol^{-1}) released at the anode of the water voltameter while 25.0 mg of copper was deposited at the cathode in the copper voltameter. (M (Cu) $= 0.063\ 54$ kg mol^{-1}).

3. A boundary between a solution of Cu^{2+} ions and one of slower colourless ions moves towards the cathode in electrolysis at a rate of 139 μm s^{-1} when the electric potential gradient in the solution is 2.50×10^3 V m^{-1}. Calculate the molar conductivity of the Cu^{2+} ion under the conditions.

4. A column of electrolyte 3.50 cm long and of cross-section 2.50 cm^2 has a resistance of 16.2 Ω. Calculate (a) the resistivity, (b) the conductivity of the electrolyte.

5. The electrolytic conductivity at 298 K of aqueous KCl containing 10.0 mol m^{-3} is 0.1413 Ω^{-1} m^{-1}. Calculate the cell constant l/a for a cell which has a resistance of 25.6 Ω at 298 K when filled with KCl of concentration 10.0 mol m^{-3}.

6. An electrolyte MX in a solution containing 5.00 mol m^{-3} has a resistance of 50.7 Ω in a conductivity cell with a cell constant of 3.50 m^{-1}. Calculate the molar conductivity of MX at that temperature and concentration. Ignore the conductivity of the water.

7. The following results show the variation of λ with c for aqueous LiCl at 298 K:

c/mol m^{-3}	0.50	1.0	5.0	10	20
$10^3\lambda/\Omega^{-1}$ m^2 mol^{-1}	11.31	11.24	10.94	10.73	10.47

By plotting λ/Ω^{-1} m^2 mol^{-1} against $(c/mol\ m^{-3})^{\frac{1}{2}}$ and extrapolating to $c = 0$, find the value of λ_0 for LiCl at 298 K.

8. For NaI, λ varies with c in aqueous solution at 298 K as follows:

$c/mol\ m^{-3}$	0.50	1.0	5.0	10	20
$10^3 \lambda/\Omega^{-1}\ m^2\ mol^{-1}$	12.54	12.43	12.13	11.92	11.67

By plotting these results as in Q. 7 above, find the value of λ_0 for NaI at 298 K.

9. Calculate λ_0 (HCO_2H) from the following data:

$$\lambda_0\ (HCl)\ \ \ \ = 42.61 \times 10^{-3}\ \Omega^{-1}\ m^2\ mol^{-1};$$
$$\lambda_0\ (NaCl)\ \ \ = 12.64 \times 10^{-3}\ \Omega^{-1}\ m^2\ mol^{-1};$$
$$\lambda_0\ (HCO_2Na)\ = 10.56 \times 10^{-3}\ \Omega^{-1}\ m^2\ mol^{-1}.$$

(Use Kohlrausch's law and solve the equation

$$\lambda_0\ (HCO_2H)\ =\ \lambda_0\ (H^+)\ +\ \lambda_0\ (HCO_2^-)$$

by the algebraic method of simultaneous equations).

10. Pure water of resistivity 2.33×10^4 Ω m was used to make a saturated solution of AgCl at 298 K. The resistivity of the solution was 6.45×10^3 Ω m. Assuming the AgCl to be so dilute that $\lambda = \lambda_0$ for the solution, make use of Table XIX to calculate the concentration of AgCl in the solution.

11. At 291 K a saturated aqueous solution of $PbSO_4$ had a resistivity of 5.43×10^2 Ω m. The resistivity of the water used to make the solution was 7.14×10^3 Ω m. The values of λ_0 for the Pb^{2+} ion and the SO_4^{2-} ion at 291 K are respectively 12.2×10^{-3} Ω^{-1} m^2 mol^{-1} and 13.6×10^{-3} Ω^{-1} m^2 mol^{-1}. Calculate the concentration of saturated $PbSO_4$ at that temperature.

12. In a moving boundary experiment in aqueous NaCl of concentration 20.0 mol m^{-3} the Na^+ ions moved 7.00 cm towards the cathode when 3.87 C were passed through the solution. The cross-section of the tube was 11.2 mm^2. Calculate t_{Na^+} for the solution.

13. The molar conductivity of an aqueous solution of MX was found to be 13.05×10^{-3} Ω^{-1} m^2 mol^{-1} at a certain concentration. The value of t_{M^+} was found to be 0.45 in the solution. Calculate $\lambda\ (M^+)$ at that concentration.

14. The resistance R of a conductivity cell containing 100.0 cm^3 of aqueous NaOH varied as shown below when a volume V

of aqueous HCl of concentration 200.0 mol m^{-3} was added.
Draw a graph of Ω/R against V/cm^3 and hence find the concentration of the original NaOH solution.

V_{HCl}/m^3	0.0	2.0	4.0	6.0	8.0	10.0	12.0	14.0	16.0
$10^{-2}R/\Omega$	2.00	2.27	2.63	3.13	3.85	5.00	4.00	2.44	1.76

15. A monobasic organic acid RCOOH at a concentration of
27.0 mol m^{-3} has a molar conductivity of $1.00 \times 10^{-3} \ \Omega^{-1} \ m^2$
mol^{-1} at 298 K. At that temperature $\lambda_0 \ (H^+) = 34.98 \times 10^{-3} \ \Omega^{-1}$
m^2 mol^{-1} and $\lambda_0 \ (R.COO^-) = 4.12 \times 10^{-3} \ \Omega^{-1} \ m^2 \ mol^{-1}$. Calculate the value of Ostwald's dissociation constant for the acid
from Eqn. (95).

16. The resistivity of an aqueous solution of an organic acid
RCOOH of concentration 14.0 mol m^{-3} was 6.40 Ω m at 298 K.
For this acid λ_0 is $39.7 \times 10^{-3} \ \Omega^{-1} \ m^2 \ mol^{-1}$ at this tempera-
ture. Calculate the conductivity ratio α at that concentration.

Chapter 13

Dipole moments and electronegativity

Polarisation of molecules in an electric field

A diatomic molecule of an element (e.g. H_2 or F_2) can be considered to have the centre of negative charge and the centre of positive charge at the mid-point of a straight line joining the atomic nuclei. However, such molecules are polarised when placed in an electric field between the plates of a condenser; the centre of negative charge is attracted to the positive plate and the centre of positive charge to the negative plate. The effect is to reduce the difference of potential U which is produced between the plates of a condenser when a charge Q is applied to one of them. The ratio $\dfrac{Q}{U}$ for a condenser is called the capacitance, C, and it is related to the dimensions of the condenser by the equation

$$C = \frac{A\epsilon}{d} \tag{96}$$

where A is the area of the two parallel plates, d is the distance between them and ϵ is the permittivity of the material between the plates.

The permittivity of a gas like hydrogen or fluorine, or in fact any substance, is greater than the permittivity of a vacuum, ϵ_0, precisely because molecules are polarised in an electric field. The ratio ϵ/ϵ_0 for a substance is called its relative permittivity, ϵ_r. Experimentally, it is the ratio between the capacitance of a condenser when that substance fills the space between the plates and the capacitance of the same condenser with a vacuum between the plates.

171

It is found experimentally that a quantity

$$\frac{\epsilon_r - 1}{\epsilon_r + 2} \times \frac{M}{\rho},$$

where M is the molar mass and ρ the density, is a constant, independent of temperature, for gases like H_2, N_2, Cl_2 and F_2. For diatomic gases like HF, HCl, ClF and NO, however, the quantity varies with temperature as shown in Fig. 37.

The reason is that these molecules have permanent dipoles. In HCl, for example, the chlorine nucleus has a some- what greater tendency to attract the electrons than does the hydrogen nucleus. The effect is that the centre of negative charge in the molecule lies rather nearer to the chlorine nucleus than does the centre of positive charge. The HCl mole- cules therefore tend to 'line up' in an electric field with the hydrogen ends towards the negative plate and the chlorine ends towards the positive. As a result of the superimposition of this effect on the polarisation caused by the field itself, the ϵ_r values of such gases as HCl and HF are high, but the effect of an increase in temperature is to reduce ϵ_r because the in- creased thermal agitation of the gas molecules (p. 47) reduces their tendency to align themselves in the field.

The gradient of the straight line in Fig. 37 can be shown to be related to N_A, the Avogadro constant, R, the gas con- stant, and ϵ_0, the permittivity of a vacuum, by the equation

$$\text{gradient} = \frac{(N_A p_e)^2}{9 \epsilon_0 R} / \text{m}^3 \text{ mol}^{-1} \text{ K}$$

in which p_e is the dipole moment of the gas molecule. For HCl the gradient is 6.52×10^{-3} m^3 mol^{-1} K.

Thus $\quad p_e \text{ (HCl)} = \dfrac{3}{N_A} (6.52 \times 10^{-3} \text{ m}^3 \text{ mol}^{-1} \text{ K} \times \epsilon_0 \times R)^{\frac{1}{2}}$

$$= 3.45 \times 10^{-30} \text{ C m}.$$

Percentage ionic character of a bond

The internuclear distance in HCl = 127.5 pm. Thus the molecule can be considered to behave, in its electrostatic properties, as two opposite charges of

$$\frac{3.45 \times 10^{-30} \text{ C m}}{127.5 \times 10^{-12} \text{ m}} = 2.7 \times 10^{-20} \text{ C}$$

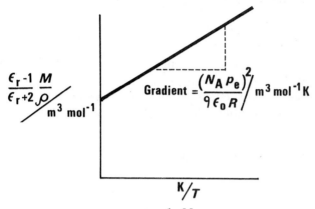

Fig. 37.   Variation of $\dfrac{\varepsilon_1 - 1}{\varepsilon_2 + 2} \cdot \dfrac{M}{\rho}$ with $1/T$ for HCl.

(about one-sixth of an electronic charge) separated by this distance.

H   Cl
· ←————————— 127.5 pm ————————→ ·

$+\dfrac{e}{6}$ $-\dfrac{e}{6}$

We say that the HCl bond has one-sixth (16.7%) ionic character.

Example

Measurements of the permittivity of ICl vapour over a range of temperature show p_e (ICl) to be 1.79×10^{-30} C m. The internuclear distance I–Cl is 231 pm. Calculate the percentage ionic character of the ICl bond.

The molecule can be represented as opposite charges of

$$\frac{1.79 \times 10^{-30} \text{ C m}}{2.31 \times 10^{-10} \text{ m}} = 7.8 \times 10^{-21} \text{ C}$$

separated by the internuclear distance. But the size of the electronic charge is 1.60×10^{-19} C.

Thus the percentage ionic character of the bond is

$$\frac{7.8 \times 10^{-21} \text{ C} \times 100}{1.60 \times 10^{-19} \text{ C}} = 4.9\%.$$

Dipole moments of polyatomic molecules

The dipole moments of some polyatomic molecules are given in Table XX.

The molecules of CO_2, BCl_3 and PCl_5 have zero dipole moments, not because their bonds are without polar character, but because the directions of the bonds are such that the polarities cancel one another. The molecule of CO_2 is linear,

$$O = C = O$$

that of BCl_3 is an equilateral triangle,

and that of PCl_5 is a trigonal bipyramid, like BCl_3 but with two additional Cl atoms, one above and one below the central atom on an axis perpendicular to the plane of the other three Cl atoms.

The table indicates, therefore, that the SO_2, H_2S and H_2O molecules are not linear and that the molecules of NH_3 and PCl_3 are not planar triangles.

Electronegativity

The electronegativity of an atom was defined by Pauling as the power of that atom within a molecule to attract electrons to itself. Pauling made use of thermochemical data to construct a table of electronegativities of elements – an electronegativity scale. For many elements, however, some of the experimental values for the required thermochemical quantities were unsatisfactory, and remain so.

In 1958 Allred and Rochow suggested another method of constructing an electronegativity scale. They based their argument on a consideration of the electrostatic attraction exerted by the nucleus of one atom on an outer electron of another atom bound to it. If r is the distance between the nucleus and that electron, the force of attraction is $Z^*e^2 / 4\pi \epsilon_0 r^2$, where Z^*e is the effective nuclear charge (p. 36) felt by the electron under the influence of that nucleus. They

TABLE XX

Dipole Moments of Some Polyatomic Molecules

	$10^{30} \, p_e/Cm$
H_2O	6.17
H_2S	3.12
NH_3	4.88
PCl_3	2.30
BCl_3	0
PCl_5	0
SO_2	5.42
CO_2	0

considered r to represent the single-bond covalent radius of the ◀━━━
atom, defined as one-half of the distance between the nuclei of
two atoms of the element which are linked by a single bond.

In order to produce an electronegativity scale similar to
that of Pauling, they plotted Z^*/r^2 (see Slater's rules, p. 36)
against Pauling's figures for the electronegativities of the elements, and obtained points grouped around the line

$$\chi = \frac{3.59 \times 10^3 \, Z^* \, pm^2}{r^2} + 0.744 \qquad (97)$$

The values of χ obtained by using this equation were used as
the electronegativities of the elements.

Example

Use Eqn. (97) to calculate the electronegativity of Cl, $1s^2$
$2s^2 \, 2p^6 \, 3s^2 \, 3p^5$, for which the single-bond covalent radius r_{Cl}
is 99 pm.

An electron adjacent to the periphery of the Cl atom is
shielded from the nucleus of that atom by

7 3s and 3p electrons, contributing 7×0.35 to s,
8 2s and 2p electrons, contributing 8×0.85,
2 1s   electrons, contributing 2×1.0.

$\therefore s$, the total screening, $= 2.0 + 6.8 + 2.45 = 11.25$

and $Z^* = Z - s = 17 - 11.25.$

$= 5.75.$

TABLE XXI
Some Values for Electronegativities of Elements
(Allred-Rochow Scale)

F	4.10	B	2.01
N	3.07	Si	1.74
Cl	2.85	Al	1.47
C	2.50	Mg	1.23
H	2.20	K	0.91

$$\therefore \chi_{Cl} = \frac{3.59 \times 10^3 \times 5.75 \text{ pm}^2}{(99 \text{ pm})^2} + 0.744$$

$$= 2.85.$$

Table XXI shows the values of electronegativities of a number of elements on the Allred-Rochow scale. It should be stressed however that these figures represent typical values; the tendency of an atom in a molecule to attract electrons depends to some extent on its environment within the molecule as well as on the position of the element in the electronegativity scale.

Electronegativities and bond polarities

The polarity of a bond $A-X$ is related to the difference in electronegativity between the elements A and X. The greater the value of $\Delta\chi$, i.e. $|\chi_A - \chi_B|$ the greater the percentage ionic character. A number of empirical equations have been suggested for the relation between bond polarity and $\Delta\chi$. The results obtained from one such scheme are shown in Table XXII.

Thus for the C-F bond $\Delta\chi = \chi(F) - \chi(C) = 4.10 - 2.50 = 1.60$ and the percentage ionic character, according to this scheme, is 47%. Such results can be tested against information on bond polarities obtained from dipole moments (p.172). Other empirical equations would give rather different relations between bond polarity and $\Delta\chi$ from those given in Table XXII, but the general principle always applies that the greater the difference in electronegativity in two atoms the greater the ionic character of the bond which joins them.

TABLE XXII

$\Delta \chi$ and Polar Character in a Bond

$\Delta \chi$	0.2	0.6	1.0	1.6	2.4	3.0
% ionic character	1	9	22	47	76	89

Examples XIII

1. The dipole moment of the HF molecule is 6.35×10^{-30} C m and the bond length is 92 pm. Calculate the percentage ionic character of the HF bond.

2. For HI $p_e = 1.39 \times 10^{-30}$ C m and the H–I distance is 162 pm. Calculate the percentage ionic character of the HI bond.

3. The single-bond covalent radius of the oxygen atom is 74 pm. Calculate χ (oxygen) on the Allred–Rochow scale. The electron configuration is $1s^2\ 2s^2\ 2p^4$.

4. Calculate χ (sulphur) from Eqn. (97). For S, the electron configuration is $1s^2\ 2s^2\ 2p^6\ 3s^2\ 3p^4$ and $r_s = 102$ pm.

5. Calculate χ (phosphorus) on the Allred–Rochow scale. $r_p = 110$ pm and P (ground state) is $1s^2\ 2s^2\ 2p^6\ 3s^2\ 3p^3$.

6. Use Table XXII to construct a smooth curve showing the relation between $\Delta \chi$ and percentage ionic character. Hence estimate the percentage ionic character of the HI bond, for which $\Delta \chi = 0.54$. Compare the result with the experimental value obtained from the dipole moment of the HI molecule (Q. 2 above).

7. Use the graph drawn in Q. 6 above to estimate the percentage ionic character of an Fe–O bond for which $\Delta \chi = 1.86$.

8. The molecule of o-dichlorobenzene has a dipole moment but its isomer p-dichlorobenzene is not polar. However, p-dihydroxybenzene has a small dipole moment. Suggest explanations for these observations.

9. The compound $PtCl_2(NH_3)_2$ exists in two isomeric forms. One has zero dipole moment but the other is polar. Suggest structures for the two compounds which would explain these observations.

Chapter 14

Titrations

Standard solutions

A standard solution is a solution of known strength, which is conveniently expressed as a concentration, c, defined by the equation:

$$c = n/V \qquad (98)$$

where n is the amount of solute contained in a volume V of solution. As the unit of amount in the SI is the mole (p. 6) and the unit of volume is the cubic metre, the unit of concentration is mol m^{-3}.

Example

An aqueous solution of anhydrous sodium carbonate, Na_2CO_3 ($M = 0.1060$ kg mol^{-1}) contains 8.36 g Na_2CO_3 in 150.0 cm^3 of solution. Calculate the concentration of the Na_2CO_3 solution.

The amount of Na_2CO_3,

$$n = \frac{8.36 \times 10^{-3} \text{ kg}}{0.1060 \text{ kg mol}^{-1}} = 7.89 \times 10^{-2} \text{ mol}$$

$$\therefore c = n/V = \frac{7.89 \times 10^{-2} \text{ mol}}{1.50 \times 10^{-4} \text{ m}^3} = 526 \text{ mol m}^{-3}.$$

The concentration of a solution can be determined experimentally if a standard solution of another substance, which will react with it quantitatively, is provided, and if a suitable indicator can be found to show the end-point. We shall be concerned here with three types of titrations:

(a) acid–base titrations,
(b) redox titrations,
(c) precipitation titrations.

The principles of acid-base titrations

An acid has a tendency, when in aqueous solution, to donate a proton to a water molecule:

$$HA + H_2O = A^- + H_3O^+,$$

whereas a base has a tendency to gain a proton from a water molecule:

$$B + H_2O = BH^+ + OH^-$$

(see also (pp. 133–138).

When an aqueous acid is added to an aqueous alkali (i.e. a water-soluble base) the H_3O^+ ions in the acid combine with the OH^- ions in the alkali:

$$H_3O^+ + OH^- = 2H_2O$$

Sodium salts of weak acids, e.g., Na_2CO_3 and $Na_2B_4O_7$, are also used as alkalis in aqueous solution because they hydrolyse to produce OH^- ions. The choice of indicator is governed by the nature of the acid and base which are to be titrated. Weak bases such as Na_2CO_3 are best titrated using an indicator like methyl orange which undergoes a colour change at low pH (p. 144). In some acid–base reactions the end-point shown by one indicator denotes a different reaction from that shown by another. In the case of Na_2CO_3 titrations, phenolphthalein becomes colourless, on the addition of acid, when the carbonate is all changed to bicarbonate:

$$Na_2CO_3 + HCl = NaHCO_3 + NaCl.$$

But the carbonate ion has to be completely converted to CO_2 before methyl orange will change from yellow to red:

e.g. $$Na_2CO_3 + 2HCl = H_2O + CO_2 + 2NaCl.$$

The importance of the indicator is further illustrated in the example which follows.

Example

Phenolphthalein was added to 25.0 cm³ of aqueous NaOH of concentration 50.0 mol m⁻³. Addition of 31.25 cm³ of H_3PO_4 of concentration 20.0 mol m⁻³ was sufficient to decolorise the phenolphthalein. Use this result to determine the stoichiometry of the reaction.

We must first use Eqn. (98) to calculate the amounts of the reactants which take part in the titration reaction.

Since $\quad c = n/V$

$\quad\quad\quad n = V \times c.$

Thus $\quad n\,(\text{NaOH}) = 25.0 \times 10^{-6}\ \text{m}^3 \times 50.0\ \text{mol m}^{-3}$

$\quad\quad\quad\quad\quad\quad = 1.250 \times 10^{-3}\ \text{mol}$

and $\quad n\,(\text{H}_3\text{PO}_4) = 31.25 \times 10^{-6}\ \text{m}^3 \times 20.0\ \text{mol m}^{-3}$

$\quad\quad\quad\quad\quad\quad = 0.625 \times 10^{-3}\ \text{mol}.$

Thus NaOH and H_3PO_4 react in the molar ratio 2:1 and the equation of reaction is

$$2\text{NaOH} + \text{H}_3\text{PO}_4 = \text{Na}_2\text{HPO}_4 + 2\text{H}_2\text{O}.$$

However, when methyl orange was used as the indicator it was found that 62.5 cm^3 of the H_3PO_4 solution were required to effect the colour change from yellow to red:

$n\,(\text{H}_3\text{PO}_4) = 62.5 \times 10^{-6}\ \text{m}^3 \times 20.0\ \text{mol m}^{-3} = 1.250 \times 10^{-3}\ \text{mol}.$

Thus NaOH and H_3PO_4 react in the molar ratio 1:1, and the equation of reaction in this titration is

$$\text{NaOH} + \text{H}_3\text{PO}_4 = \text{NaH}_2\text{PO}_4 + \text{H}_2\text{O}.$$

Primary standards: standardisation of solutions

There are comparatively few acids or bases which can be weighed out accurately for the purpose of making a standard solution. The requirements for such a primary standard are that it should be (a) easily purified and (b) non-hygroscopic. A good primary standard for acid-base titrations is anhydrous sodium carbonate, which is best made by prolonged heating of the bicarbonate. The following example illustrates the use of Na_2CO_3 for the standardisation of an acid.

Example

Anhydrous Na_2CO_3 (1.535 g) was made up to 250.0 cm^3 with distilled water. In a titration with methyl orange as indicator, 21.8 cm^3 of aqueous HCl were needed to neutralise 25.0 cm^3 of the standard carbonate solution. Calculate the concentration of the HCl. $M\,(\text{Na}_2\text{CO}_3) = 0.1060\ \text{kg mol}^{-1}.$

The equation of the reaction is

$$Na_2CO_3 + 2HCl = 2NaCl + H_2O + CO_2.$$

Thus 1 mole Na_2CO_3 requires 2 moles HCl.

25.0 cm^3 of Na_2CO_3 solution contains $\dfrac{25.0 \text{ cm}^3}{250.0 \text{ cm}^3} \times 1.535$ g Na_2CO_3

$$= \frac{1}{10} \times \frac{1.535 \times 10^{-3} \text{ kg}}{0.1060 \text{ kg mol}^{-1}}$$

$$= 1.448 \times 10^{-3} \text{ mol } Na_2CO_3$$

From the equation of reaction, the amount of HCl required to neutralise the Na_2CO_3 is $2 \times 1.448 \times 10^{-3}$ mol. Since this is contained in 21.8 cm^3, the concentration of the HCl solution is

$$\frac{2 \times 1.448 \times 10^{-3} \text{ mol}}{21.8 \times 10^{-6} \text{ m}^3} = 133 \text{ mol m}^{-3}.$$

Back-titration

Some reactions do not proceed satisfactorily to completion unless one of the reactants is in excess. Thus to determine the amount of acid required to dissolve a certain amount of metal, a standard acid is added in excess and the amount remaining at the end of the reaction is determined by titration against a standard alkali. The procedure is known as a back-titration. The principle is illustrated by the example which follows.

Example

An ammonium salt (2.60 g) was boiled with 70.0 cm^3 of an NaOH solution until all the ammonia was expelled:

$$NH_4^+ + NaOH = NH_3 + Na^+ + H_2O.$$

The remaining solution was made up to 100.0 cm^3 with distilled water, and the average of several titrations showed that 20.0 cm^3 of this solution required 16.8 cm^3 of an H_2SO_4 solution containing 250.0 mol m^{-3} H_2SO_4. If 10.0 cm^3 of the original NaOH solution required 21.0 cm^3 of the same H_2SO_4 solution, calculate the percentage of NH_4^+ ion in the ammonium salt.

Let us calculate first the amount of NaOH added to the ammonium salt.

10.0 cm^3 of the original NaOH solution required 21.0 cm^3 of aqueous sulphuric acid containing 250.0 mol m^{-3} H_2SO_4, i.e. 21.0 × 10^{-6} m^3 × 250.0 mol m^{-3} = 5.25 × 10^{-3} mol. Since the titration reaction is

$$2NaOH + H_2SO_4 = Na_2SO_4 + 2H_2O,$$

it follows that 10.0 cm^3 of the original NaOH solution contained 2 × 5.25 × 10^{-3} mol NaOH = 1.05 × 10^{-2} mol. Therefore the original 70.0 cm^3 added to the ammonium salt contained 7 × 1.05 × 10^{-2} mol = 7.35 × 10^{-2} mol.

Let us now calculate the amount of NaOH which remains after boiling the original 70.0 cm^3 with the ammonium salt. One-fifth of this NaOH (20 cm^3 from a prepared 100 cm^3 of solution) required 16.8 cm^3 of the same sulphuric acid solution,

i.e. 16.8 × 10^{-6} m^3 × 250.0 mol m^{-3} = 4.20 × 10^{-3} mol H_2SO_4

Thus all the remaining NaOH would have required 5 × 4.20 × 10^{-3} mol H_2SO_4, and, from the equation of reaction, this represents 2 × 5 × 4.20 × 10^{-3} mol NaOH = 4.20 × 10^{-2} mol. Therefore the amount of NaOH used in the reaction

$$NH_4^+ + NaOH = NH_3 + Na^+ + H_2O$$

was (7.35−4.20) × 10^{-2} mol = 3.15 × 10^{-2} mol NaOH, and the ammonium salt contained 3.15 × 10^{-2} mol NH_4^+.

Since M (NH_4^+) = 0.0180 kg mol^{-1}, the weight of NH_4^+ in the salt is

$$3.15 \times 10^{-2} \text{ mol} \times 0.0180 \text{ kg mol}^{-1} = 0.567 \times 10^{-3} \text{ kg}$$

$$= 0.567 \text{ g.}$$

∴ The percentage of NH_4^+ ion in the salt is

$$\frac{0.567 \text{ g}}{2.60 \text{ g}} \times 100 = 21.8\%.$$

The principles of redox titrations

The formal concept of <u>oxidation number</u> is useful in the balancing of redox equations, particularly those containing oxo-anions. The basic rules for the allocation of an oxidation number to an element are enumerated below.

1. The free element has zero oxidation number.

2. The oxidation number of the element in a monatomic ion is equal to the charge on the ion. Thus oxygen has the

oxidation number -2 in the O^{2-} ion and iron has the oxidation number $+3$ in the Fe^{3+} ion.

3. In a molecule or a polyatomic ion the oxidation number of a particular element is the charge its atom might be expected to carry if the molecule were composed entirely of ions, the anions among them being assumed to have noble gas structure. Thus in SO_2 the sulphur has an oxidation number of $+4$. Since oxygen is more electronegative than sulphur, a formal division into ions would produce two O^{2-} ions with the electronic structure of neon, and one S^{4+} ion. In the $SO_4{}^{2-}$ ion the sulphur has an oxidation number $+6$; a formal division into ions would produce four O^{2-} ions and therefore one S^{6+} ion; the sum of the oxidation numbers of the atoms must equal the total charge on the species.

The application of oxidation number to balancing redox equations can be illustrated by the reduction of chlorate to chloride. In the $ClO_3{}^-$ ion the chlorine is in the $+5$ state whereas in Cl^- it is in the -1 state. Thus six electrons are required for the reduction of $ClO_3{}^-$ to Cl^-. In acid solution the half-equation is

$$ClO_3{}^- + 6e + 6H^+ = Cl^- + 3H_2O.$$

In a half-equation for the reduction of oxoanions in acid solution it is usually necessary to balance the charge on the two sides by adding H^+ ions to one side and H_2O molecules to the other. In this case six H^+ ions on the left and three H_2O molecules on the right are necessary to ensure that the total charge and the stoichiometry are both balanced.

As the oxidation of Fe^{2+} to Fe^{3+} releases only one electron:

$$Fe^{2+} = Fe^{3+} + e,$$

six Fe^{2+} ions are necessary to reduce one $ClO_3{}^-$ ion to Cl^-. The complete redox equation is obtained by adding six times the second half-equation to the first half-equation:

$$\begin{array}{l} ClO_3{}^- + 6H^+ + 6e = Cl^- + 3H_2O \\ \underline{\phantom{ClO_3{}^- + }6Fe^{2+} = 6Fe^{3+} + 6e} \\ ClO_3{}^- + 6H^+ + 6Fe^{2+} = Cl^- + 3H_2O + 6Fe^{3+}. \end{array}$$

Note that there must be no electron in the balanced equation.

Potassium permanganate titrations

Potassium permanganate is a particularly useful oxidising agent in acidic solution. On reduction, the highly coloured MnO_4^- ion is converted to the almost colourless Mn^{2+} ion, therefore no indicator is required:

$$MnO_4^- + 8H^+ + 5e = Mn^{2+} + 4H_2O.$$

Permanganate is a strong oxidising agent (p. 151). It is usually acidified with sulphuric acid; HCl is unsuitable because it is oxidised to chlorine by MnO_4^- (Table XVIII). We shall illustrate the use of potassium permanganate by examples of the oxidation of Fe^{2+} to Fe^{3+} and of oxalic acid to CO_2. In the latter case the titration has to be carried out at about 340 K, otherwise the reaction is very slow.

Example A

A specimen of iron wire weighing 0.2545 g was dissolved in dilute H_2SO_4 and made up to 100.0 cm^3 with distilled water. This solution (20.0 cm^3) required 18.1 cm^3 of a solution of $KMnO_4$ containing 10.0 mol m^{-3} for titration. Calculate the percentage of iron in the wire. $M(Fe^{2+}) = 0.055\ 85$ kg mol^{-1}.

Since the half-reactions are

$$MnO_4^- + 8H^+ + 5e = Mn^{2+} + 4H_2O$$
and $\qquad Fe^{2+} \qquad\qquad = Fe^{3+} + e,$

the balanced reaction is

$$MnO_4^- + 8H^+ + 5Fe^{2+} = Mn^{2+} + 5Fe^{3+} + 4H_2O.$$

18.1 cm^3 of $KMnO_4$ of concentration 10.0 mol m^{-3} contains

$$18.1 \times 10^{-6}\ m^3 \times 10.0\ mol\ m^{-3}$$

$$= 1.81 \times 10^{-4}\ mol\ KMnO_4$$

$\therefore$ 20 cm^3 of the iron solution contains (from the equation of reaction)

$$5 \times 1.81 \times 10^{-4}\ mol = 9.05 \times 10^{-4}\ mol\ Fe.$$

As only one-fifth of the iron solution was used in the titration, the amount of iron in the original wire was

$$5 \times 9.05 \times 10^{-4}\ mol\ Fe$$

$$= 4.525 \times 10^{-3}\ mol\ Fe$$

$$= 4.525 \times 10^{-3}\ mol \times 0.055\ 85\ kg\ mol^{-1}\ Fe$$

$$= 0.2527 \text{ g Fe.}$$

$$\therefore \% \text{ Fe in wire} = \frac{0.2527 \text{ g} \times 100}{0.2545\text{g}} = 99.3\%.$$

Example B

10.4 g of a solution saturated with oxalic acid ($H_2C_2O_4$) at 298 K was made up to 200 cm^3 with distilled water. The mean result for titration of a 20.0 cm^3 portion of this solution with potassium permanganate (concentration 20.0 mol m^{-3}) was 18.4 cm^3. Calculate the solubility of oxalic acid at 298 K.

The oxidation number of carbon in the $C_2O_4^{2-}$ ion, and therefore in $H_2C_2O_4$, is +3, whereas in CO_2 it is +4. Thus two electrons must be removed from $H_2C_2O_4$ to convert it to two CO_2 molecules.

$$H_2C_2O_4 = 2CO_2 + 2e + 2H^+$$

As the half-equation for the reduction of MnO_4^- is

$$MnO_4^- + 8H^+ + 5e = Mn^{2+} + 4H_2O,$$

the balanced equation involves the transference of 10 electrons, and is

$$5H_2C_2O_4 + 2MnO_4^- + 6H^+ = 2Mn^{2+} + 10CO_2 + 8H_2O.$$

As 18.4 cm^3 of the KMnO$_4$ solution contains

$$18.4 \times 10^{-6} \text{ m}^3 \times 20.0 \text{ mol m}^{-3} = 3.68 \times 10^{-4} \text{ mol}$$

$\therefore$ 20 cm^3 of the prepared $H_2C_2O_4$ solution contains

$$\frac{5}{2} \times 3.68 \times 10^{-4} \text{ mol} = 9.2 \times 10^{-4} \text{ mol } H_2C_2O_4$$

(from the equation).

Thus the original 200 cm^3 of prepared oxalic acid solution contained

$$10 \times 9.2 \times 10^{-4} \text{ mol} = 9.2 \times 10^{-3} \text{ mol } H_2C_2O_4$$
$$= 9.2 \times 10^{-3} \text{ mol} \times 0.0900 \text{ kg mol}^{-1}$$
$$= 0.828 \text{ g.}$$

Thus 10.4 g solution contained 0.828 g $H_2C_2O_4$. The solubility is therefore

$$\frac{0.828 \text{ g}}{(10.4 - 0.828) \text{ g}} = 8.65 \times 10^{-2}$$

(See p. 130 for the definition of solubility).

Iodine titrations

Iodine dissolves in aqueous KI to give a brown solution containing I_3^- ions. The strength of the solution can be determined with a standard solution of sodium thiosulphate pentahydrate, $Na_2S_2O_3.5H_2O$. The reaction is

$$2 \ S_2O_3^{2-} + I_2 = 2 \ I^- + S_4O_6^{2-}.$$

The colour of an iodine solution is weakened as standard thiosulphate is added to it. Near the end-point, when the solution has a pale straw-colour, a little starch paste is added. A deep blue colour appears which is destroyed at the end-point when sufficient $S_2O_3^{2-}$ ions are added to convert the last of the iodine to iodide.

Example

Sodium thiosulphate pentahydrate (2.065 g) was dissolved and made up to 250.0 cm^3 with distilled water. In the titration of 20.0 cm^3 of a solution of iodine in aqueous KI, 18.4 cm^3 of the standard thiosulphate solution was used. Calculate the concentration of I_2 in the iodine solution. $M \ (Na_2S_2O_3.5H_2O)$ = 0.2480 kg mol^{-1}.

The amount of $Na_2S_2O_3.5H_2O$ used in one titration is

$$\frac{2.065 \times 10^{-3} \text{ kg}}{0.2480 \text{ kg mol}^{-1}} \times \frac{18.4 \text{ cm}^3}{250.0 \text{ cm}^3} = 6.13 \times 10^{-4} \text{ mol.}$$

Therefore, from the equation of reaction, n_{I_2} in 20.0 cm^3 is

$$\frac{6.13 \times 10^{-4} \text{ mol}}{2}$$

$$\therefore c_{I_2} = \frac{n_{I_2}}{V} = \frac{6.13 \times 10^{-4} \text{ mol}}{2 \times 20.0 \times 10^{-6} \text{ m}^3}$$

$$= 15.3 \text{ mol m}^{-3}.$$

Volumetric methods based on the oxidation of aqueous KI to I_2

A number of substances can be determined in solution by using them to oxidise KI solution and then titrating with thiosulphate the iodine which is produced. Examples are:

$$2\ Cu^{2+} + 4\ I^- = 2\ CuI + I_2$$
$$Cl_2 + 2\ I^- = 2\ Cl^- + I_2$$
$$6\ H^+ + IO_3^- + 5\ I^- = 3\ H_2O + 3\ I_2$$

Weakly acidic conditions are necessary in all these reactions, but too low a pH causes the $Na_2S_2O_3$ to decompose.

Example

An excess of potassium iodide was added to 25.0 cm^3 of a solution of Cu^{2+} ions in slightly acidic solution. For the titration of the iodine which was released it was necessary to add 21.4 cm^3 of a solution of $Na_2S_2O_3$ containing 50.0 mol m^{-3}. Calculate the concentration of the original Cu^{2+} solution.

The amount of $Na_2S_2O_3$ used in the titration is

$$21.4 \times 10^{-6}\ m^3 \times 50.0\ mol\ m^{-3} = 1.07 \times 10^{-3}\ mol.$$

But, from the equations of reaction:

$$2\ Cu^{2+} + 4\ I^- = 2\ CuI + I_2$$
and
$$I_2 + 2\ S_2O_3^{2-} = S_4O_6^{2-} + 2\ I^-,$$

it is seen that one mole of Cu^{2+} requires one mole of $S_2O_3^{2-}$. Thus 25.0 cm^3 of the Cu^{2+} contains 1.07×10^{-3} mol and

$$c_{Cu^{2+}} = \frac{1.07 \times 10^{-3}\ mol}{2.50 \times 10^{-5}\ m^3} = 42.8\ mol\ m^{-3}.$$

Precipitation analysis

Chloride and bromide ions can be determined in neutral solution with standard aqueous silver nitrate. About 1 cm^3 of 5% potassium chromate solution is used as indicator. Addition of $AgNO_3$ causes precipitation of the halide.

$$AgNO_3 + Cl^- = AgCl + NO_3^-$$

Not until all the halide ion is precipitated does the formation of the reddish-brown Ag_2CrO_4 begin:

$$2\ AgNO_3 + CrO_4^{2-} = Ag_2CrO_4 + 2\ NO_3^-$$

This is the Mohr titration for determining halide. It is not satisfactory for iodide, for which an adsorption indicator is preferable, or for acidic solutions. In the latter case the Volhard method can be used. This involves a back-titration (p. 181). An excess of $AgNO_3$ is added to the halide solution, about 1 cm^3 of saturated aqueous ferric alum is introduced, and standard potassium thiocyanate is added from a burette. The Ag^+ ions react according to the equation:

$$CNS^- + Ag^+ = AgCNS.$$

When all the Ag^+ ions are removed from solution the addition of further CNS^- produces the strong blood-red coloration of the $FeCNS^{2+}$ ion which indicates the end-point.

Example

To 25.0 cm^3 of an NaCl solution was added an $AgNO_3$ solution (50.0 cm^3) containing 110.0 mol m^{-3}. A precipitate of AgCl was produced, and the excess of Ag^+ ions required 10.50 cm^3 of a KCNS solution containing 98.0 mol m^{-3}. Calculate the amount of NaCl in the original solution and hence the chloride ion concentration.

The amount of $AgNO_3$ added to the chloride solution was:

$$50.0 \times 10^{-6} \ m^3 \times 110.0 \ mol \ m^{-3} = 5.500 \times 10^{-3} \ mol.$$

The amount of KCNS needed for the back-titration was:

$$10.5 \times 10^{-6} \ m^3 \times 98.0 \ mol \ m^{-3} = 10.29 \times 10^{-4} \ mol.$$

Thus the amount of Ag^+ ion needed to precipitate the chloride was:

$$(5.500 - 1.029) \times 10^{-3} \ mol$$
$$= 4.471 \times 10^{-3} \ mol.$$

Thus, from the equation:

$$Ag^+ + Cl^- = AgCl,$$

the original 25.0 cm^3 of chloride solution contained 4.471×10^{-3} mol, and the concentration of chloride ion was:

$$\frac{4.471 \times 10^{-3} \ mol}{25.0 \times 10^{-6} \ m^3} = 1.788 \times 10^2 \ mol \ m^{-3}.$$

Another precipitation reaction which can be used for a volumetric determination is that between Ag^+ ions and CN^- ions. If aqueous silver nitrate is added to a neutral solution of a cyanide the soluble $Ag(CN)_2^-$ ion is formed at first:

$$Ag^+ + 2 \ CN^- = Ag(CN)_2^-.$$

When all the CN^- ions are converted to $Ag(CN)_2^-$ ions, addition of further Ag^+ produces slight turbidity because AgCN is insoluble

$$Ag(CN)_2^- + Ag^+ = 2\ AgCN.$$

The end-point, marked by the appearance of turbidity, therefore denotes the completion of the first reaction above.

Examples XIV

(For atomic weights see Table XXV, p. 206)

1. Calculate the concentration of a solution of Na_2CO_3, 20.0 cm^3 of which required 22.8 cm^3 of a solution of HCl containing 375 mol m^{-3} for neutralisation when methyl orange was used as indicator.

2. 16.9 cm^3 of a solution of oxalic acid containing 23.4 g of $H_2C_2O_4.2H_2O$ crystals in a total volume of 1000.0 cm^3 were required to neutralise 25.0 cm^3 of an NaOH solution when phenol-phthalein was used as indicator. Calculate the concentration of the NaOH solution. The reaction between oxalic acid and NaOH under these conditions is

$$2\ NaOH + H_2C_2O_4 = Na_2C_2O_4 + 2\ H_2O.$$

3. Sodium borate solution can be titrated with aqueous HCl using methyl orange as indicator, the reaction being

$$Na_2B_4O_7 + 2\ HCl + 5\ H_2O = 2\ NaCl + 4\ H_3BO_3.$$

Crystalline sodium borate, $Na_2B_4O_7, xH_2O$ (31.5 g) was dissolved in water and made up to 1000.0 cm^3. 25.0 cm^3 of the solution required 11.8 cm^3 of a solution of HCl of concentration 350.0 mol m^{-3} for titration with methyl orange as indicator. Calculate the molar mass of the crystalline solid and hence x in the formula above.

4. A solution containing both Na_2CO_3 and $NaHCO_3$ (25.0 cm^3) required 11.36 cm^3 HCl of concentration 110.0 mol m^{-3} when phenolphthalein was used as indicator but a further 39.77 cm^3 of the same acid when methyl orange was used. Calculate the concentration of (a) Na_2CO_3 (b) $NaHCO_3$ in the solution.

5. An aqueous solution containing both HCl and H_3PO_4 (25.0 cm^3) required 19.25 cm^3 of a solution of NaOH of concentration 100.0 mol m^{-3} for neutralisation when methyl orange was used as an indicator, but a further 5.25 cm^3 of the same NaOH solution when phenolphthalein was used (see p. 179). Calculate the concentration of (a) H_3PO_4 (b) HCl in the solution.

6. A mixture of NH_4Cl and NaCl (200.0 mg in all) was dissolved in 25.0 cm^3 of aqueous NaOH containing 100.0 mol m^{-3}. After

all the NH_3 had been driven off by boiling (see p. 181) it was found that 12.0 cm^3 of a solution of HCl containing 50.0 mol m^{-3} were needed to neutralise the remaining NaOH. Calculate the weights of NH_4Cl and NaCl in the original mixture.

7. What is the oxidation number of sulphur in the following molecules and ions? (a) SO_3 (b) H_2S (c) SCl_2 (d) S_2Cl_2 (e) SO_3^{2-} (f) $S_2O_3^{2-}$ (g) $S_4O_6^{2-}$.

8. Pure $Fe(NH_4SO_4)_2.6H_2O$ crystals (1.560 g) required 40.7 cm^3 of a $KMnO_4$ solution for oxidation. Calculate the concentration of the $KMnO_4$ solution.

9. An aqueous solution X contains both Fe^{2+} and Fe^{3+} ions. 25.0 cm^3 of X required for oxidation 20.0 cm^3 of a solution of $KMnO_4$ containing 20.0 mol m^{-3}. A further 25 cm^3 of X were reduced with amalgamated zinc to convert Fe^{3+} to Fe^{2+} and then required 32.0 cm^3 of the same $KMnO_4$ solution. Calculate the concentrations of both Fe^{2+} and Fe^{3+} ions in X.

10. The salt $KH_3(C_2O_4)_2$, potassium trihydrogen dioxalate, was dissolved in water, and 20.0 cm^3 of the solution required 12.0 cm^3 of NaOH of concentration 500 mol m^{-3} to neutralise it. A further 20.0 cm^3 of the solution required 40.0 cm^3 of $KMnO_4$ solution to oxidise the oxalate ions to CO_2. Calculate the concentration of the $KMnO_4$ solution.

11. A solution contained $H_2C_2O_4$ and H_2SO_4; a 20.0 cm^3 portion required 19.1 cm^3 of a $KMnO_4$ solution containing 20.0 mol m^{-3} for oxidation and another 20.0 cm^3 portion required 28.0 cm^3 of an NaOH solution containing 220.0 mol m^{-3} for neutralisation. Calculate the concentrations of $H_2C_2O_4$ and H_2SO_4 in the original solution.

12. An aqueous solution of H_2O_2 (50.0 cm^3) was made acidic, and an excess of KI was added:

$$2\ I^- + H_2O_2 + 2\ H^+ = I_2 + 2\ H_2O.$$

The iodine which was liberated required 20.0 cm^3 of a solution of $Na_2S_2O_3$ containing 100.0 mol m^{-3}. Calculate the concentration of the H_2O_2 solution.

13. A specimen of bleaching powder (1.70 g) was ground to a fine paste with water and made up to 200.0 cm^3 with aqueous KI and acetic acid. A 20.0 cm^3 portion of the solution required 18.5 cm^3 of a solution of $Na_2S_2O_3$ containing 80.0 mol m^{-3} in order to reduce the iodine liberated by the chlorine in the

bleaching powder. Calculate the percentage of available chlorine in the powder. M (Cl_2) = 0.070 92 kg mol^{-1}.

14. To 20.0 cm^3 of a solution of KIO_3 was added an excess of KI. The solution was acidified, and the iodine which was liberated needed 24.5 cm^3 of $Na_2S_2O_3$ containing 75.0 mol m^{-3}. Calculate the concentration of the original KIO_3 solution.

15. A solution X contains NaCl and HCl. The total Cl$^-$ ion was determined by adding 20.0 cm^3 of a solution of $AgNO_3$ containing 100.0 mol m^{-3} to 20.0 cm^3 of X and back-titrating with 6.0 cm^3 of KCNS of concentration 80.0 mol m^{-3}. A further 20.0 cm^3 of X required 8.5 cm^3 of an NaOH solution containing 100.0 mol m^{-3} for neutralisation. Calculate the concentrations of NaCl and HCl in X.

16. A mixture of NaCl and KBr crystals (1.89 g) was dissolved and made up to 200.0 cm^3 with water. A 20.0 cm^3 portion needed 22.0 cm^3 of an $AgNO_3$ solution containing 100.0 mol m^{-3} in a titration with chromate as indicator. Calculate the weight of NaCl in the original mixture.

17. To 100.0 cm^3 of a freshly prepared solution of H_2S in water, 50.0 cm^3 of aqueous $AgNO_3$ of concentration 100.0 mol m^{-3} were added:

$$H_2S + 2\ AgNO_3 = Ag_2S + 2\ HNO_3.$$

The precipitate was filtered off and washed. The filtrate and washings required 39.0 cm^3 of aqueous NaOH containing 100.0 mol m^{-3} for titration with methyl orange. Calculate the concentration of the original H_2S solution.

18. Calculate the volume of aqueous KCNS of concentration 40.0 mol m^{-3} which would have been needed to back-titrate the excess of Ag$^+$ ions after filtration of the Ag_2S in Q. 17 above.

19. Potassium chlorate in aqueous solution was completely reduced to chloride using SO_2:

$$ClO_3^- + 3\ SO_2 + 3\ H_2O = Cl^- + 3\ SO_4^{2-} + 6\ H^+.$$

The solution was boiled to remove the excess of SO_2, and 50.0 cm^3 of $AgNO_3$ solution containing 100.0 mol m^{-3} were added. The excess of Ag$^+$ ion required 20.5 cm^3 of KCNS containing 80.0 mol m^{-3} for back-titration. Calculate the amount of $KClO_3$ in the original solution.

20. A sample containing only iron and Fe_2O_3 (225.0 mg in total) was dissolved in acid and all the iron was brought to the oxidation

state +2. 37.5 cm^3 of aqueous $KMnO_4$ containing 19.82 mol m^{-3} were needed to oxidise all the Fe^{2+} to Fe^{3+}. Calculate the percentages of Fe and Fe_2O_3 in the sample.

Chapter 15

Quantitative analysis of organic compounds

We shall use the elemental analysis of organic compounds to illustrate some general aspects of quantitative analysis. The principles of calculation for inorganic analysis are essentially the same.

A. Elemental analysis of solids and liquids

Analysis for carbon and hydrogen

A weighed quantity of an organic compound is heated in a stream of dry oxygen or CO_2-free air, and the vapour passes through hot CuO to complete the oxidation of hydrogen to H_2O vapour and carbon to CO_2. The gases are passed first through tubes of dry magnesium perchlorate which absorbs H_2O and then through tubes containing asbestos impregnated with KOH which absorbs CO_2.

Since 1 mole C ($M = 0.0120$ kg mol^{-1}) $\rightarrow$ 1 mole CO_2 ($M = 0.0440$ kg mol^{-1}), a sample which produces a weight w of CO_2 (i.e. the increase in weight of the KOH tube) contains a weight

$$\frac{0.0120 \text{ kg mol}^{-1}}{0.0440 \text{ kg mol}^{-1}} w \;=\; \frac{3.0\, w}{11.0} \text{ of carbon.}$$

Similarly, H_2 ($M = 0.0020$ kg mol^{-1}) $\rightarrow$ H_2O ($M = 0.0180$ kg mol^{-1}), and the weight of hydrogen in the original sample is

$$\frac{0.0020 \text{ kg mol}^{-1}}{0.0180 \text{ kg mol}^{-1}} w' \;=\; \frac{1.0\, w'}{9.0},$$

where w' is the increase in the weight of the magnesium per-chlorate tube.

Example

A sample of an organic compound (14.08 mg) was completely oxidised to H_2O (7.92 mg) and CO_2 (48.40 mg). Calculate the percentages of hydrogen and carbon in the compound.

$$\%H = \frac{1.0}{9.0} \times \frac{7.92 \text{ mg}}{14.08 \text{ mg}} \times 100 = 6.25\%.$$

$$\%C = \frac{3.0}{11.0} \times \frac{48.40 \text{ mg}}{14.08 \text{ mg}} \times 100 = 93.8\%.$$

Analysis for nitrogen: Dumas' method

A weighed quantity of a nitrogen-containing compound is heated with an excess of CuO in an atmosphere of CO_2. The vapour is passed over a hot copper spiral to reduce any oxides of nitrogen to nitrogen, which is collected in a graduated tube over aqueous KOH which absorbs the CO_2.

Example

An organic compound (51.0 mg) produced, in analysis, 6.08 cm^3 of nitrogen gas, measured at 290 K and 1.011×10^5 N m^{-2}. Calculate the percentage of nitrogen in the compound.

Assuming nitrogen to behave ideally:

$$pV = nRT = \frac{w}{M} RT$$

and $\quad w(N_2) = \frac{pVM}{RT}$

$$= \frac{1.011 \times 10^5 \text{ N m}^{-2} \times 6.08 \times 10^{-6} \text{ m}^3 \times 0.0280 \text{ kg mol}^{-1}}{8.314 \text{ J K}^{-1} \text{ mol}^{-1} \times 290 \text{ K}}$$

$$= 7.14 \times 10^{-6} \text{ kg} = 7.14 \text{ mg}.$$

$$\therefore \% \text{ nitrogen in the compound} = \frac{7.14 \text{ mg} \times 100}{51.0 \text{ mg}}$$

$$= 14.0\%.$$

Analysis for nitrogen: Kjeldahl's method

In most nitrogen compounds the element can be converted to the ammonium ion by digesting the compound with a little concentrated sulphuric acid containing potassium sulphate, to raise the boiling point, and mercury, to act as catalyst. The mixture is boiled for about thirty minutes, cooled and diluted. An excess

· of sodium hydroxide solution is added and the ammonia is steam-distilled into standard acid. The acid is then back-titrated with standard alkali to determine the amount of ammonia distilled over.

Example

An organic nitrogen compound (83.5 mg) was digested with sulphuric acid to convert the nitrogen to NH_4^+. When the reaction was complete the liquid was made alkaline with sodium hydroxide. The ammonia which was distilled off on boiling was passed into 50 cm^3 of an HCl solution containing 50.0 mol m^{-3}. The acid then required 35.0 cm^3 of standard NaOH containing 50.0 mol m^{-3} for back-titration. Calculate the percentage of nitrogen in the original compound.

The amount of HCl in the solution originally

$$= 50.0 \times 10^{-6} \ m^3 \times 50.0 \ mol \ m^{-3} = 2.50 \times 10^{-3} \ mol.$$

The amount of HCl left after absorption of the NH_3

$$= 35.0 \times 10^{-6} \ m^3 \times 50.0 \ mol \ m^{-3} = 1.75 \times 10^{-3} \ mol.$$

Since the reaction between NH_3 and HCl is

$$NH_3 + HCl = NH_4Cl,$$

the amount of NH_3 absorbed is $(2.50 - 1.75) \times 10^{-3}$ mol

$$= 0.75 \times 10^{-3} \ mol.$$

One mole NH_3 contains 0.0140 kg.

$\therefore$ 0.75×10^{-3} mol contains $0.0140 \times 0.75 \times 10^{-3}$ kg

$$= 10.5 \ mg.$$

The percentage of nitrogen in the original compound was therefore

$$\frac{10.5 \ mg}{83.5 \ mg} \times 100 = 12.6\%.$$

Analysis for sulphur

The oxygen-flask technique is used for sulphur analysis. A weighed amount of the compound is wrapped in filter paper which is placed in a small platinum-gauze cage attached to the ground-glass stopper of a large conical flask. The paper is ignited, and the sample is immediately introduced into the flask, which contains oxygen and about 20 cm^3 of water containing a little H_2O_2.

The SO_2 formed in the combustion is converted to H_2SO_4 by the H_2O_2 and is titrated with standard alkali.

Example

A sample weighing 68.3 mg, combusted in an oxygen-flask, gave an aqueous solution of H_2SO_4 which required 21.5 cm^3 of an NaOH solution of concentration 20.0 mol m^{-3} for titration. Calculate the percentage of sulphur in the compound.

21.5×10^{-6} m$^3 \times 20.0$ mol m^{-3} NaOH $= 4.30 \times 10^{-4}$ mol neutralises 2.15×10^{-4} mol H_2SO_4, since the equation of neutralisation is

$$2 \text{ NaOH} + H_2SO_4 = Na_2SO_4 + 2 H_2O.$$

This amount of H_2SO_4 contains 2.15×10^{-4} mol S

$$= 2.15 \times 10^{-4} \text{ mol} \times 0.0320 \text{ kg mol}^{-1}$$

$$= 6.88 \times 10^{-6} \text{ kg} = 6.88 \text{ mg}.$$

$$\therefore \text{ \%S in compound} = \frac{6.88 \text{ mg} \times 100}{68.3 \text{ mg}} = 10.1\%.$$

Analysis for chlorine and bromine

The oxygen-flask technique used for determination of chlorine or bromine is like that used for sulphur except that the HCl or HBr which is formed is usually absorbed in aqueous NaOH. The solution is made exactly neutral with HNO_3 and titrated against $AgNO_3$ with chromate as indicator (p. 187).

Example

A sample weighing 53.6 mg, combusted in an oxygen-flask, gave a solution of chloride ions ($M = 0.03545$ kg mol^{-1}) which needed for titration 10.8 cm^3 of a solution of $AgNO_3$ containing 50.0 mol m^{-3}. Calculate the percentage of chlorine in the compound.

10.8×10^{-6} m$^3 \times 50.0$ mol m$^{-3} = 5.40 \times 10^{-4}$ mol $AgNO_3$ is needed to titrate 5.40×10^{-4} mol Cl$^-$, since the titration equation is

$$\text{Cl}^- + AgNO_3 = AgCl + NO_3^-.$$

5.40×10^{-4} mol Cl$^- = 5.40 \times 10^{-4}$ mol $\times 0.03545$ kg mol^{-1}

$$= 1.92 \times 10^{-5} \text{ kg} = 19.2 \text{ mg}.$$

$$\therefore \quad \%Cl \text{ in compound} = \frac{19.2 \text{ mg} \times 100}{53.6 \text{ mg}} = 35.7\%.$$

Analysis for oxygen

The amount of oxygen in a compound is determined by difference.

Example

An organic compound containing carbon, hydrogen, chlorine and oxygen gave the following results on analysis.

A sample (33.1 mg) gave 17.6 mg CO_2 and 5.40 mg H_2O on combustion. Another sample (46.3 mg) produced, in an oxygen-flask combustion, chloride ion, which required for titration 16.78 cm^3 of a solution of $AgNO_3$ containing 50.0 mol m^{-3}. Calculate the percentages of C, H, Cl and O in the compound.

$$\%C = \frac{17.6 \text{ mg} \times 3.0 \times 100}{33.1 \text{ mg} \times 11.0} = 14.50\%.$$

$$\%H = \frac{5.40 \text{ mg} \times 1.0 \times 100}{33.1 \text{ mg} \times 9.0} = 1.81\%.$$

$$\%Cl = \frac{16.78 \times 10^{-6} \text{ m}^3 \times 50.0 \text{ mol m}^{-3} \times 0.03545 \text{ kg mol}^{-1} \times 100}{46.3 \times 10^{-6} \text{ kg}}$$

$$= 64.25\%.$$

$$\therefore \%O \text{ (by difference)} = 100.00 - 14.50 - 1.81 - 64.25 = 19.44\%.$$

Calculation of empirical formula

The empirical formula of a compound is obtained by dividing the percentage of each element by the corresponding relative atomic mass to obtain the ratio of atoms. In the case above:

For carbon $\qquad \dfrac{14.50}{12.00} = 1.21$

For hydrogen $\qquad \dfrac{1.81}{1.01} = 1.79$

For chlorine $\qquad \dfrac{64.25}{35.45} = 1.80$

For oxygen $\qquad \dfrac{19.44}{16.00} = 1.21.$

Dividing through by the highest common factor, 0.60, the ratio of atoms, $C : H : Cl : O$ is $2 : 3 : 3 : 2$.

Thus the simplest formula (the empirical formula) is $C_2H_3Cl_3O_2$.

Calculation of molecular formula

For the compound above the molar mass was found by a cryoscopic determination (p. 120) to be $0.1655 \text{ kg mol}^{-1}$. The molar mass of the compound $(C_2H_3Cl_3O_2)_n$, by addition of atomic masses, is $0.1655\, n \text{ kg mol}^{-1}$. Thus $n = 1$ and the molecular formula is $C_2H_3Cl_3O_2$.

B. Analysis of gaseous hydrocarbons and mixtures

Determining the molecular formula of a gaseous hydrocarbon

A gas of molecular formula C_xH_y, on combustion with an excess of oxygen, will react:

$$C_xH_y \text{ (g)} + (x + \frac{y}{4}) \ O_2 \text{ (g)} = xCO_2 \text{ (g)} + \frac{y}{2} \ H_2O \text{ (l)}$$

Thus the amount of gas is reduced by $(1 + \frac{y}{4})$ mol for every mole of reaction. If the CO_2 is absorbed in aqueous potash the amount of gas is further reduced by x mol. If the pressure at the end of the combustion is equal to that at the beginning, the amounts of gas are proportional to the volumes (Avogadro's law, p. 41).

Example

A gaseous hydrocarbon C_xH_y (10.0 cm^3) was mixed with an excess of oxygen (50.0 cm^3) and sparked to ignite the gas. The volume of the residual gas was 40.0 cm^3, of which 20.0 cm^3 were absorbed on treatment with aqueous potash. All measurements were made at the same pressure. Deduce the molecular formula of the gas.

From the equation above, reduction in volume/volume of hydrocarbon

$$= \frac{(10.0 + 50.0 - 40.0) \text{ cm}^3}{10.0 \text{ cm}^3} = 1 + \frac{1}{4}y.$$

$$\therefore \ y = 4.$$

Also, reduction in volume on adding KOH/volume of hydrocarbon

$$= \frac{20.0 \text{ cm}^3}{10.0 \text{ cm}^3} = x = 2.$$

Therefore the molecular formula of the hydrocarbon is C_2H_4.

The composition of gaseous mixtures

Similar use is made of Avogadro's law in the determination of the composition of a gaseous mixture of hydrocarbons.

Example

A mixture containing CH_4, C_2H_2 and C_2H_4 (27.0 cm³) was sparked with an excess (75.0 cm³) of oxygen. The residual gas occupied 52.5 cm³. A solution of KOH absorbed 45.0 cm³ of that gas. Calculate the composition of the mixture.

Let the volumes of CH_4, C_2H_2 and C_2H_4 be respectively x, y and z.

Reduction in volume
on sparking

(i) $\quad CH_4 \ + 2 \ O_2 \ = \quad CO_2 + 2 \ H_2O \ (l)$
$\quad\quad$ x $\quad\quad$ 2x $\quad\quad\quad$ x $\quad\quad\quad\quad\quad\quad\quad$ 2x

(ii) $\quad C_2H_2 + 2\frac{1}{2}O_2 = 2 \ CO_2 + \quad H_2O \ (l)$
$\quad\quad$ y $\quad\quad$ 2.5y $\quad\quad$ 2y $\quad\quad\quad\quad\quad\quad$ 1.5y

(iii) $\quad C_2H_4 + 3 \ O_2 = 2 \ CO_2 + 2 \ H_2O \ (l)$
$\quad\quad$ z $\quad\quad$ 3z $\quad\quad\quad$ 2z $\quad\quad\quad\quad\quad\quad$ 2z

(a) Original volume $\quad = x + y + z = 27.0 \text{ cm}^3$.

(b) Reduction on
$\quad$ sparking $\quad\quad = 2x + 1.5y + 2z = (27.0 + 75.0 - 52.5) \text{ cm}^3$
$\quad\quad\quad\quad\quad\quad\quad\quad\quad\quad\quad\quad = 49.5 \text{ cm}^3$.

(c) CO_2 dissolved
$\quad$ by KOH $\quad\quad = x + 2y + 2z = 45.0 \text{ cm}^3$

Subtracting (c) from 2 × (a):

$$x = 9.0 \text{ cm}^3.$$

Substituting in (b) and (c)

$$1.5y + 2z = 31.5 \text{ cm}^3$$
$$2y + 2z = 36.0 \text{ cm}^3.$$

Subtracting: $\quad\quad\quad$ 0.5y $\quad = 4.5 \text{ cm}^3$
$\quad\quad\quad\quad\quad\quad\quad \therefore \ y \quad = 9.0 \text{ cm}^3$
$\quad\quad \therefore \text{ from (a) } z \quad = 9.0 \text{ cm}^3$.

Thus the gas contains 9.0 cm^3 of CH_4, 9.0 cm^3 of C_2H_2 and 9.0 cm^3 of C_2H_4.

Examples XV

(For atomic weights see Table XXV)

1. An organic compound containing only C, H and O, of weight 203.6 mg, gave on complete combustion 389.5 mg CO_2 and 239.0 mg H_2O. Calculate the empirical formula of the compound.

2. Calculate the empirical formula of a compound containing only C, H, N and S from the following results:

    (a) 57.0 mg on complete combustion gave 143.0 mg CO_2 and 27.0 mg H_2O.

    (b) 91.2 mg gave 9.74 cm^3 N_2 at 293 K and 1.013×10^5 N m^{-2}.

    (c) 76.0 mg, oxidised in an oxygen flask, gave H_2SO_4 which required 6.68×10^{-4} mol NaOH for neutralisation.

3. An organic liquid containing only C, H and O, of weight 20.0 mg, gave on complete combustion 36.2 mg CO_2 and 12.3 mg H_2O. Calculate the empirical formula.

4. A gaseous hydrocarbon (9.0 cm^3), on combustion with 60.0 cm^3 of oxygen, gave a gas which occupied 42.0 cm^3 at the same temperature and pressure. Addition of aqueous KOH reduced the volume to 15 cm^3 of oxygen. Calculate the molecular formula of the hydrocarbon.

5. A sample of a nitrogen-containing compound (103.5 mg) was digested with sulphuric acid to convert all the nitrogen to NH_4^+. When an excess of NaOH was added and the ammonia was steam-distilled into 40.0 cm^3 of HCl containing 50.0 mol m^{-3}, the acid needed 16.0 cm^3 of NaOH containing 50.0 mol m^{-3} for back-titration. Calculate the percentage of nitrogen in the compound.

6. A liquid with the composition C, 66.7; H, 11.1; O, 22.2% reacts with 2:4-dinitrophenylhydrazine to give a solid with the composition C, 47.6; H, 4.8; N, 22.2; O, 25.4%. The liquid reacts with ammoniacal silver nitrate on warming to produce a precipitate of silver. Suggest two possible structures for the liquid compound.

7. A colourless organic solid of composition C, 80.0; H, 6.7; O, 13.3% reacts with hydroxylamine to give a solid of composition C, 71.3; H, 6.7; N, 10.3; O, 11.7%. The former compound does not reduce ammoniacal silver nitrate; write down its structural formula.

8. An organic acid crystallises in a hydrated form which has the composition C, 34.3; H, 4.7; O, 61.0%. A sample of the hydrated acid (70.0 mg) neutralised 20.0 cm^3 of NaOH containing 50.0 mol m^{-3}. A specimen of the ethyl ester of the acid was found in a cryoscopic determination to have a molar mass of 0.276 kg mol^{-1}. Deduce the molecular formula of the anhydrous acid.

9. An organic liquid of composition C, 77.3; H, 7.6; N, 15.1%, on treatment with acetic anhydride, gave a solid of composition C, 71.0; H, 6.7; N, 10.4; O, 11.9%, which was found by a cryoscopic determination to have a molar mass of 0.135 kg mol^{-1}. Deduce the formula of the liquid.

10. An aliphatic compound, A, contained C, 21.2; H, 1.8; Cl, 62.8%. When it was treated with water in the cold the product, B, contained C, 25.4; H, 3.15 and Cl, 37.6%. When B was treated with sodium carbonate solution and subsequently acidified, an acid, C, of molecular formula $C_2H_4O_3$ was isolated. Write structural formulae for A, B and C.

11. A solid X contained C, 49.3; H, 9.6; N, 19.2%. When the solid was heated with P_2O_5 a liquid, Y, of composition C, 65.5; H, 9.1; N, 25.4% was obtained. Hot aqueous sodium hydroxide set free all the nitrogen from Y in the form of ammonia. The alkaline solution was acidified with HCl and an organic acid, Z, of molecular formula $C_3H_6O_2$, was isolated from it. Write down structural formulae for X, Y and Z.

12. A neutral compound, A, of empirical formula $C_8H_7O_2$, was found by an ebullioscopic determination to have a molar mass of about 0.27 kg mol^{-1}. When A was boiled with aqueous KOH a clear solution was obtained which, when saturated with CO_2, gave a liquid, B, which turned to a low-melting solid of molecular formula C_6H_6O. The compound B gave a purple colour with aqueous $FeCl_3$ and a white precipitate with bromine water. The aqueous solution from which B was removed gave, on acidification with HCl, a solid organic acid, C, of empirical formula $C_2H_3O_2$, which was converted to a neutral compound, D, when it was refluxed with methanol and a trace of mineral

acid. The compound **D** had an empirical formula $C_3H_5O_2$ and a density of 6.6 kg m^{-3} at 1.01×10^5 N m^{-2} and 275 K. When **C** is heated it loses water and the product is a neutral compound **E**, with the molecular formula $C_4H_4O_3$. Deduce the structures of **A**, **B**, **C**, **D** and **E**.

13. A liquid **X** of composition C, 38.8; H, 9.7; O, 51.5% reacted with acetyl chloride to give a compound **Y** of composition C, 50.0; H, 5.6; O, 44.4% and molar mass 0.146 kg mol^{-1}. What are the compounds **X** and **Y**?

14. An organic liquid had the composition C, 61.3; H, 5.1; N, 10.2; O, 23.4%. On reduction it was converted to a base, the hydrochloride of which contained 24.7% Cl. When the base was treated with $NaNO_2$ and acid at 273 K and subsequently with KCN and $CuSO_4$, a compound of molecular formula C_8H_7N was produced which was hydrolysed to p-toluic acid on boiling with dilute H_2SO_4. Formulate the original compound and write equations for the reactions described.

15. A compound **X** of composition C, 20.0; H, 6.6; N, 46.7; O, 26.7% was converted by gentle heating into a compound of empirical formula $C_2O_2N_3H_5$. When **X** was treated with HNO_2 the only products of the reaction were CO_2, H_2O and N_2. Write a structural formula for **X** and equations for the reactions described.

16. An organic liquid of composition C, 49.3; H, 6.9; O, 43.8% has a molar mass of 0.146 kg mol^{-1}. It reacts with ammonia to give a white precipitate with the empirical formula $CONH_2$. This solid reacts with aqueous NaOH to give NH_3 and an aqueous solution of a sodium salt which on evaporating to dryness and treating with concentrated H_2SO_4 gives a mixture of CO_2 and CO. Write the formula for the liquid and equations for the reactions described.

17. A mixture of CH_4, CO and N_2, of total volume 13.8 cm^3, was exploded with an excess of oxygen. The observed contraction in volume (measured at the original temperature) was 13.4 cm^3 and there was a further contraction of 11.2 cm^3 on treatment with aqueous KOH. Calculate the composition of the original mixture.

18. A pure gaseous hydrocarbon (20.0 cm^3) was exploded with oxygen (120.0 cm^3). A contraction in volume of 60.0 cm^3 occurred and there was a further contraction of 60.0 cm^3 on treatment with aqueous alkali. All volumes were measured

under the same conditions of temperature and pressure. What is the molecular formula of the hydrocarbon?

19. A mixture of CH_4, C_2H_4 and C_2H_2 of total volume 15.2 cm^3 was mixed with an excess of oxygen and sparked. The contraction in volume was 26.6 cm^3 and there was a further contraction of 26.6 cm^3 when the gas was treated with aqueous alkali. The volumes were all measured under the same conditions. Calculate the composition of the mixture.

20. A mixture of C_2H_4 and C_3H_8 (20.0 cm^3) was mixed with 100.0 cm^3 of oxygen and exploded. The volume of residual gas, measured at the same temperature and pressure, was 68.6 cm^3. After absorption with aqueous alkali the volume was reduced to 17.2 cm^3. Calculate the percentages of C_2H_4 and C_3H_8 in the mixture.

Appendix

TABLE XXIII
Values of Some Physical Constants
(4 Significant Figures)

Physical constant	Symbol	Value	Logarithm of the measure
speed of light in a vacuum	c_O	2.998×10^8 m s^{-1}	8.4769
permeability of a vacuum	μ_O	$4\pi \times 10^{-7}$ kg m s^{-2} A^{-2}	$\overline{6}.0992$
permittivity of a vacuum	ϵ_O	8.854×10^{-12} kg^{-1} m^{-3} s^4 A^2	$\overline{12}.9471$
unified atomic mass constant	m_u	1.660×10^{-27} kg	$\overline{27}.2201$
mass of electron	m_e	9.109×10^{-31} kg	$\overline{31}.9594$
electronic charge	e	-1.602×10^{-19} C	$\overline{19}.2046$
Planck constant	h	6.626×10^{-34} J s	$\overline{34}.8213$
Avogadro constant	N_A	6.023×10^{23} mol^{-1}	23.7798
gas constant	R	8.314 J K^{-1} mol^{-1}	0.9198
Faraday constant	F	9.649×10^4 C mol^{-1}	4.9845
ice-point temperature	T_{ice}	273.2 K	2.4365

TABLE XXIV
Some non-SI Units Which are Widely Used

Physical quantity	Unit	Symbol	Definition in terms of SI units
Length:	ångström	Å	10^{-10} m
	micron	μ	10^{-6} m
Volume:	litre	l	10^{-3} m^3
Energy:	kilowatt hour	kWh	3.6×10^6 J
	thermochemical calorie	cal	4.184 J
	erg	erg	10^{-7} J
	electronvolt	eV	1.6021×10^{-19} J*
Pressure:	atmosphere	atm	101 325 N m^{-2}

* Not a true unit: best experimental value.

TABLE XXV
The Periodic Table

H 1.008 1																	He 4.003 2
Li 6.94 3	Be 9.01 4											B 10.81 5	C 12.01 6	N 14.01 7	O 16.00 8	F 19.00 9	Ne 20.18 10
Na 22.99 11	Mg 24.31 12											Al 26.98 13	Si 28.09 14	P 30.97 15	S 32.06 16	Cl 35.45 17	Ar 39.95 18
K 39.10 19	Ca 40.08 20	Sc 44.96 21	Ti 47.90 22	V 50.94 23	Cr 52.00 24	Mn 54.94 25	Fe 55.85 26	Co 58.93 27	Ni 58.71 28	Cu 63.55 29	Zn 65.37 30	Ga 69.72 31	Ge 72.59 32	As 74.92 33	Se 78.96 34	Br 79.90 35	Kr 83.80 36
Rb 85.47 37	Sr 87.62 38	Y 88.91 39	Zr 91.22 40	Nb 92.91 41	Mo 95.94 42	Tc 43	Ru 101.07 44	Rh 102.91 45	Pd 106.4 46	Ag 107.87 47	Cd 112.40 48	In 114.82 49	Sn 118.69 50	Sb 121.75 51	Te 127.60 52	I 126.90 53	Xe 131.30 54
Cs 132.91 55	Ba 137.34 56	La 138.91 57	Hf 178.49 72	Ta 180.95 73	W 183.85 74	Re 186.2 75	Os 190.2 76	Ir 192.2 77	Pt 195.1 78	Au 197.0 79	Hg 200.6 80	Tl 204.4 81	Pb 207.2 82	Bi 209.0 83	Po 84	At 85	Rn 86
Fr 87	Ra 88	Ac 89															

Lanthanides

Ce 58	Pr 59	Nd 60	Pm 61	Sm 62	Eu 63	Gd 64	Tb 65	Dy 66	Ho 67	Er 68	Tm 69	Yb 70	Lu 71

Actinides

Th 90	Pa 91	U 92	Np 93	Pu 94	Am 95	Cm 96	Bk 97	Cf 98	Es 99	Fm 100	Md 101	No 102	Lr 103

TABLE XXVI

Greek Letters Used in the Text

alpha	α	mu	μ
beta	β	nu	ν
gamma	γ	pi	π
delta	Δ, δ	rho	ρ
epsilon	ϵ	sigma	σ
theta	θ	chi	χ
kappa	κ	psi	ψ
lambda	λ	omega	Ω, ω

Note on molality and molarity

The molality, m, of a solution is defined as amount of solute divided by mass of solvent. Thus a solution of 0.20 mol of solute in 1.00 kg of solvent has a molality of 0.20 mol kg^{-1}.

The concentration, c, of a solution is expressed in the SI in mol m^{-3}. At 277 K, in a very dilute aqueous solution in which the mass of the solute itself is very small compared with the mass of water,

$$c/\text{mol m}^{-3} = 10^3 \, m/\text{mol kg}^{-1}$$

because 1.0 m^3 H$_2$O at 277 K weighs 10^3 kg.

The word molarity is not used in the SI. It was formerly used to mean amount of solute divided by total volume of solution, and expressed in the units mol l^{-1}. As 1 l = 10^{-3} m^3:

$$c/\text{mol m}^{-3} = 10^3 \times molarity/\text{mol l}^{-1}.$$

	0	1	2	3	4	5	6	7	8	9	1	2	3	4	5	6	7	8
10	·0000	0043	0086	0128	0170	0212	0253	0294	0334	0374	4	8	12	17	21	25	29	33
11	·0414	0453	0492	0531	0569	0607	0645	0682	0719	0755	4	8	11	15	19	23	26	30
12	·0792	0828	0864	0899	0934	0969	1004	1038	1072	1106	3	7	10	14	17	21	24	28
13	·1139	1173	1206	1239	1271	1303	1335	1367	1399	1430	3	6	10	13	16	19	23	26
14	·1461	1492	1523	1553	1584	1614	1644	1673	1703	1732	3	6	9	12	15	18	21	24
15	·1761	1790	1818	1847	1875	1903	1931	1959	1987	2014	3	6	8	11	14	17	20	22
16	·2041	2068	2095	2122	2148	2175	2201	2227	2253	2279	3	5	8	11	13	16	18	21
17	·2304	2330	2355	2380	2405	2430	2455	2480	2504	2529	2	5	7	10	12	15	17	20
18	·2553	2577	2601	2625	2648	2672	2695	2718	2742	2765	2	5	7	9	12	14	16	19
19	·2788	2810	2833	2856	2878	2900	2923	2945	2967	2989	2	4	7	9	11	13	16	18
20	·3010	3032	3054	3075	3096	3118	3139	3160	3181	3201	2	4	6	8	11	13	15	17
21	·3222	3243	3263	3284	3304	3324	3345	3365	3385	3404	2	4	6	8	10	12	14	16
22	·3424	3444	3464	3483	3502	3522	3541	3560	3579	3598	2	4	6	8	10	12	14	15
23	·3617	3636	3655	3674	3692	3711	3729	3747	3766	3784	2	4	6	7	9	11	13	15
24	·3802	3820	3838	3856	3874	3892	3909	3927	3945	3962	2	4	5	7	9	11	12	14
25	·3979	3997	4014	4031	4048	4065	4082	4099	4116	4133	2	3	5	7	9	10	12	14
26	·4150	4166	4183	4200	4216	4232	4249	4265	4281	4298	2	3	5	7	8	10	11	13
27	·4314	4330	4346	4362	4378	4393	4409	4425	4440	4456	2	3	5	6	8	9	11	13
28	·4472	4487	4502	4518	4533	4548	4564	4579	4594	4609	2	3	5	6	8	9	11	12
29	·4624	4639	4654	4669	4683	4698	4713	4728	4742	4757	1	3	4	6	7	9	10	12
30	·4771	4786	4800	4814	4829	4843	4857	4871	4886	4900	1	3	4	6	7	9	10	11
31	·4914	4928	4942	4955	4969	4983	4997	5011	5024	5038	1	3	4	6	7	8	10	11
32	·5051	5065	5079	5092	5105	5119	5132	5145	5159	5172	1	3	4	5	7	8	9	11
33	·5185	5198	5211	5224	5237	5250	5263	5276	5289	5302	1	3	4	5	6	8	9	10
34	·5315	5328	5340	5353	5366	5378	5391	5403	5416	5428	1	3	4	5	6	8	9	10
35	·5441	5453	5465	5478	5490	5502	5514	5527	5539	5551	1	2	4	5	6	7	9	10
36	·5563	5575	5587	5599	5611	5623	5635	5647	5658	5670	1	2	4	5	6	7	8	10
37	·5682	5694	5705	5717	5729	5740	5752	5763	5775	5786	1	2	3	5	6	7	8	9
38	·5798	5809	5821	5832	5843	5855	5866	5877	5888	5899	1	2	3	5	6	7	8	9
39	·5911	5922	5933	5944	5955	5966	5977	5988	5999	6010	1	2	3	4	5	7	8	
40	·6021	6031	6042	6053	6064	6075	6085	6096	6107	6117	1	2	3	4	5	6	8	
41	·6128	6138	6149	6160	6170	6180	6191	6201	6212	6222	1	2	3	4	5	6	7	
42	·6232	6243	6253	6263	6274	6284	6294	6304	6314	6325	1	2	3	4	5	6	7	
43	·6335	6345	6355	6365	6375	6385	6395	6405	6415	6425	1	2	3	4	5	6	7	
44	·6435	6444	6454	6464	6474	6484	6493	6503	6513	6522	1	2	3	4	5	6	7	
45	·6532	6542	6551	6561	6571	6580	6590	6599	6609	6618	1	2	3	4	5	6	7	
46	·6628	6637	6646	6656	6665	6675	6684	6693	6702	6712	1	2	3	4	5	6	7	
47	·6721	6730	6739	6749	6758	6767	6776	6785	6794	6803	1	2	3	4	5	5	6	
48	·6812	6821	6830	6839	6848	6857	6866	6875	6884	6893	1	2	3	4	4	5	6	
49	·6902	6911	6920	6928	6937	6946	6955	6964	6972	6981	1	2	3	4	4	5	6	
50	·6990	6998	7007	7016	7024	7033	7042	7050	7059	7067	1	2	3	3	4	5	6	
51	·7076	7084	7093	7101	7110	7118	7126	7135	7143	7152	1	2	3	3	4	5	6	
52	·7160	7168	7177	7185	7193	7202	7210	7218	7226	7235	1	2	2	3	4	5	6	
53	·7243	7251	7259	7267	7275	7284	7292	7300	7308	7316	1	2	2	3	4	5	6	
54	·7324	7332	7340	7348	7356	7364	7372	7380	7388	7396	1	2	2	3	4	5	6	
	0	1	2	3	4	5	6	7	8	9	1	2	3	4	5	6	7	

O	1	2	3	4	5	6	7	8	9	1	2	3	4	5	6	7	8	9
·7404	7412	7419	7427	7435	7443	7451	7459	7466	7474	1	2	2	3	4	5	5	6	7
·7482	7490	7497	7505	7513	7520	7528	7536	7543	7551	1	2	2	3	4	5	5	6	7
·7559	7566	7574	7582	7589	7597	7604	7612	7619	7627	1	2	2	3	4	5	5	6	7
·7634	7642	7649	7657	7664	7672	7679	7686	7694	7701	1	1	2	3	4	4	5	6	7
·7709	7716	7723	7731	7738	7745	7752	7760	7767	7774	1	1	2	3	4	4	5	6	7
·7782	7789	7796	7803	7810	7818	7825	7832	7839	7846	1	1	2	3	4	4	5	6	6
·7853	7860	7868	7875	7882	7889	7896	7903	7910	7917	1	1	2	3	4	4	5	6	6
·7924	7931	7938	7945	7952	7959	7966	7973	7980	7987	1	1	2	3	3	4	5	6	6
·7993	8000	8007	8014	8021	8028	8035	8041	8048	8055	1	1	2	3	3	4	5	5	6
·8062	8069	8075	8082	8089	8096	8102	8109	8116	8122	1	1	2	3	3	4	5	5	6
·8129	8136	8142	8149	8156	8162	8169	8176	8182	8189	1	1	2	3	3	4	5	5	6
·8195	8202	8209	8215	8222	8228	8235	8241	8248	8254	1	1	2	3	3	4	5	5	5
·8261	8267	8274	8280	8287	8293	8299	8306	8312	8319	1	1	2	3	3	4	5	5	6
·8325	8331	8338	8344	8351	8357	8363	8370	8376	8382	1	1	2	3	3	4	4	5	6
·8388	8395	8401	8407	8414	8420	8426	8432	8439	8445	1	1	2	2	3	4	4	5	6
·8451	8457	8463	8470	8476	8482	8488	8494	8500	8506	1	1	2	2	3	4	4	5	6
·8513	8519	8525	8531	8537	8543	8549	8555	8561	8567	1	1	2	2	3	4	4	5	5
·8573	8579	8585	8591	8597	8603	8609	8615	8621	8627	1	1	2	2	3	4	4	5	5
·8633	8639	8645	8651	8657	8663	8669	8675	8681	8686	1	1	2	2	3	4	4	5	5
·8692	8698	8704	8710	8716	8722	8727	8733	8739	8745	1	1	2	2	3	4	4	5	5
·8751	8756	8762	8768	8774	8779	8785	8791	8797	8802	1	1	2	2	3	3	4	5	5
·8808	8814	8820	8825	8831	8837	8842	8848	8854	8859	1	1	2	2	3	3	4	5	5
·8865	8871	8876	8882	8887	8893	8899	8904	8910	8915	1	1	2	2	3	3	4	4	5
·8921	8927	8932	8938	8943	8949	8954	8960	8965	8971	1	1	2	2	3	3	4	4	5
·8976	8982	8987	8993	8998	9004	9009	9015	9020	9025	1	1	2	2	3	3	4	4	5
·9031	9036	9042	9047	9053	9058	9063	9069	9074	9079	1	1	2	2	3	3	4	4	5
·9085	9090	9096	9101	9106	9112	9117	9122	9128	9133	1	1	2	2	3	3	4	4	5
·9138	9143	9149	9154	9159	9165	9170	9175	9180	9186	1	1	2	2	3	3	4	4	5
·9191	9196	9201	9206	9212	9217	9222	9227	9232	9238	1	1	2	2	3	3	4	4	5
·9243	9248	9253	9258	9263	9269	9274	9279	9284	9289	1	1	2	2	3	3	4	4	5
·9294	9299	9304	9309	9315	9320	9325	9330	9335	9340	1	1	2	2	3	3	4	4	5
·9345	9350	9355	9360	9365	9370	9375	9380	9385	9390	1	1	1	2	3	3	4	4	5
·9395	9400	9405	9410	9415	9420	9425	9430	9435	9440	0	1	1	2	2	3	3	4	4
·9445	9450	9455	9460	9465	9469	9474	9479	9484	9489	0	1	1	2	2	3	3	4	4
·9494	9499	9504	9509	9513	9518	9523	9528	9533	9538	0	1	1	2	2	3	3	4	4
·9542	9547	9552	9557	9562	9566	9571	9576	9581	9586	0	1	1	2	2	3	3	4	4
·9590	9595	9600	9605	9609	9614	9619	9624	9628	9633	0	1	1	2	2	3	3	4	4
·9638	9643	9647	9652	9657	9661	9666	9671	9675	9680	0	1	1	2	2	3	3	4	4
·9685	9689	9694	9699	9703	9708	9713	9717	9722	9727	0	1	1	2	2	3	3	4	4
·9731	9736	9741	9745	9750	9754	9759	9763	9768	9773	0	1	1	2	2	3	3	4	4
·9777	9782	9786	9791	9795	9800	9805	9809	9814	9818	0	1	1	2	2	3	3	4	4
·9823	9827	9832	9836	9841	9845	9850	9854	9859	9863	0	1	1	2	2	3	3	4	4
·9868	9872	9877	9881	9886	9890	9894	9899	9903	9908	0	1	1	2	2	3	3	4	4
·9912	9917	9921	9926	9930	9934	9939	9943	9948	9952	0	1	1	2	2	3	3	4	4
·9956	9961	9965	9969	9974	9978	9983	9987	9991	9996	0	1	1	2	2	3	3	3	4
O	1	2	3	4	5	6	7	8	9	1	2	3	4	5	6	7	8	9

ANSWERS

Examples I

1. (a) mass $\times$ length^2 $\times$ time^{-3} $\times$ current^{-2}
   (b) mass^{-1} $\times$ length^{-2} $\times$ time^4 $\times$ current^2
   (c) mass $\times$ length^2 $\times$ time^{-1}

2. (a) kg m^2 s^{-3} A^{-2}
   (b) A^2 s^4 kg^{-1} m^{-2}
   (c) J A^{-1}
   (d) V s

3. (a) V
   (b) s
   (c) m

4. (a) 13.265 nm
   (b) 0.235 kg
   (c) 1.08×10^4 mV
   (d) 1.745 8 kJ
   (e) 163.5 pF

5. $F = 3.2 \times 10^{-5}$ N

6. $V = 1.00$ m^3

7. $\lambda = 3.72 \times 10^{-3}$ s^{-1}

Examples II

1. A_r (Mg) $= 24.313$

2. A_r (Si) $= 28.088$

3. (a) $B.E.$ (^{12}C) $= 1.478 \times 10^{-11}$ J, $B.E/n = 1.232 \times 10^{-12}$ J/n
   (b) $B.E.$ (^{13}C) $= 1.557 \times 10^{-11}$ J, $B.E/n = 1.197 \times 10^{-12}$ J/n

4. (a) $B.E.$ (^{10}B) $= 1.038 \times 10^{-11}$ J, $B.E/n = 1.038 \times 10^{-12}$ J/n
   (b) $B.E.$ (^{11}B) $= 1.221 \times 10^{-11}$ J, $B.E/n = 1.110 \times 10^{-12}$ J/n

5. (a) $B.E/n$ (^{27}Al) $= 1.367 \times 10^{-12}$ J/n
    (b) $B.E/n$ (^{64}Cu) $= 1.425 \times 10^{-12}$ J/n

6. (a) $^{27}_{13}\text{Al} + ^{1}_{0}\text{n} \longrightarrow \ ^{4}_{2}\text{He} + ^{24}_{11}\text{Na}$

    (b) $^{9}_{4}\text{Be} + ^{4}_{2}\text{He} \longrightarrow \ ^{12}_{6}\text{C} + ^{1}_{0}\text{n}$

    (c) $^{35}_{16}\text{S} \longrightarrow \ ^{0}_{-1}\beta + ^{35}_{17}\text{Cl}$

7. (a) $E = 6.03 \times 10^{10}$ J mol^{-1}
    (b) $E = 4.63 \times 10^{11}$ J mol^{-1}

8. (a) $\lambda = 4.84 \times 10^{-3}$ s^{-1}
    (b) $t_{\frac{1}{2}} = 143.2$ s

9.   Age $= 1150$ y

Examples III

1. (a) 1.95×10^{16} Hz
    (b) 5.12×10^{14} Hz
    (c) 6.14×10^{12} Hz
    (d) 1.01×10^{6} Hz

2. (a) 7.26×10^{7} m^{-1}
    (b) 1.58×10^{6} m^{-1}
    (c) 3.58×10^{-3} m^{-1}

3. $h\nu_0 = 6.24 \times 10^{-19}$ J

4. $E = 1.53 \times 10^{-19}$ J

5. (a) 3.12×10^{-19} J
    (b) 1.06×10^{-17} J
    (c) 3.38×10^{-21} J

6. (a) $\lambda = 123$ pm
    (b) $\lambda = 388$ pm
    (c) $\lambda = 1.74$ nm

7. $\lambda = 4.05$ pm

8. (a) $\lambda = 397$ nm
    (b) $\lambda = 93.77$ nm
    (c) $\lambda = 955$ nm

9. $\Delta E = 1.55 \times 10^{-19}$ J

212   SI UNITS IN CHEMISTRY

10. (a) $s = 8.8$
    (b) $s = 2.75$
    (c) $s = 4.15$

11. (a) $Z^* = 2.30$ e
    (b) $Z^* = 3.20$ e

Examples IV

1. $V = 4.27 \times 10^{-3} \text{ m}^3$

2. $p = 1.62 \times 10^5 \text{ N m}^{-2}$

3. $V = 5.25 \times 10^{-4} \text{ m}^3$

4. $T = 498 \text{ K}$

5. $V = 2.14 \times 10^{-4} \text{ m}^3$

6. $T = 278 \text{ K}$

7. $V = 3.64 \times 10^{-3} \text{ m}^3$

8. $n = 5.11 \times 10^{-2} \text{ mol}$

9. $M = 8.44 \times 10^{-2} \text{ kg mol}^{-1}$

10. $M = 1.61 \times 10^{-2} \text{ kg mol}^{-1}$

11. 78.2 s

12. (a) $p_A = 1.52 \times 10^4 \text{ N m}^{-2}$,
       $p_B = 2.42 \times 10^4 \text{ N m}^{-2}$
    (b) $p = 3.94 \times 10^4 \text{ N m}^{-2}$

13. $\bar{c} = 1.84 \times 10^3 \text{ m s}^{-1}$

14. $\bar{c} = 661 \text{ m s}^{-1}$

15. (a) $p = 3.80 \times 10^6 \text{ N m}^{-2}$
    (b) $p = 5.67 \times 10^6 \text{ N m}^{-2}$

16. (a) $p = 8.26 \times 10^5 \text{ N m}^{-2}$
    (b) $p = 8.31 \times 10^5 \text{ N m}^{-2}$

17. $r_{He} = 133 \text{ pm}$

18. $T_c = 310 \text{ K}, \ p_c = 5.05 \times 10^6 \text{ N m}^{-2}$
    $V_c = 1.91 \times 10^{-4} \text{ m}^3 \text{ mol}^{-1}$

19. $T_c = 5.2 \text{ K}, \ p_c = 2.27 \times 10^5 \text{ N m}^{-2}$
    $V_c = 7.11 \times 10^{-5} \text{ m}^3 \text{ mol}^{-1}$

20. $b = 3.22 \times 10^{-5} \text{ m}^3 \text{ mol}^{-1}$
    $a = 0.140 \text{ N m}^4 \text{ mol}^{-2}$

Examples V

1. $d = 316$ pm

2. $d = 205$ pm

3. $F = 5.02 \times 10^{-9}$ N

4. $E = 2.64 \times 10^{-18}$ J

5. $U = 3.55$ MJ mol^{-1}

6. $U = 4.13$ MJ mol^{-1}

7. $0.154\ r$

8. $0.732\ r$

9. $0.318\ r,\ r_c/r_a = 0.225$

10. $r_{K^+} = 133$ pm, $r_{Cl^-} = 181$ pm

11. $r_{Rb^+} = 152$ pm, $r_{Br^-} = 196$ pm

12. $r_{K^+} + r_{Br^-} = 329$ pm, $r_{Rb^+} + r_{Cl^-} = 333$ pm

13. $r_{Tl^+} = 158$ pm

14. $N_A = 6.04 \times 10^{23}$ mol^{-1}

15. $N_A = 5.99 \times 10^{23}$ mol^{-1}

16. $U = 5.964$ MJ mol^{-1}

17. $\Delta H° = +748$ kJ mol^{-1}

18. $A_{Cl} = 360$ kJ mol^{-1}

19. $\Delta H° = -342$ kJ mol^{-1}

20. D (C–F) $= +472$ kJ mol^{-1}

Examples VI

1. $\Delta H = 3.23$ MJ mol^{-1}

2. $\Delta H = 57.3$ kJ mol^{-1}

3. $\Delta H = 48.1$ kJ mol^{-1}

4. $\Delta H = 2.89$ MJ mol^{-1}

5. (a) $H°$ (n-C$_5$H$_{12}$ gas) $= -146$ kJ mol^{-1}
    (b) $H°$ (H$_2$C$_2$O$_4$) $\quad = -925$ kJ mol^{-1}
    (c) $H°$ ((NH$_2$)$_2$CO) $\quad = -333$ kJ mol^{-1}

6. $\Delta H° = -128$ kJ mol^{-1}

7. $\Delta H^\circ \quad = -310$ kJ mol^{-1}

8. $\Delta nRT \quad = +1.24$ kJ mol^{-1}

9. E (C–H) $= 412$ kJ mol^{-1}

10. E (C–C) $= 347$ kJ mol^{-1}

11. E (C≡C) $= 815$ kJ mol^{-1}

12. (a) E (N–H) $= 390$ kJ mol^{-1}
    (b) E (N–N) $= 205$ kJ mol^{-1}

13. E (C–Cl) $\quad = 328$ kJ mol^{-1}

14. E (C=O) $\quad = 690$ kJ mol^{-1}

15. E (C–Br) $\quad = 278$ kJ mol^{-1}

16. Resonance stabilisation energy $= 154$ kJ mol^{-1}

17. H° by calculation $= -278$ kJ mol^{-1}
    Resonance stabilisation energy $= \ \ 85$ kJ mol^{-1}

18. $H^\circ \ = -425$ kJ mol^{-1}

19. $\Delta H^\circ = -635$ kJ mol^{-1}

20. $H^\circ \ = + \ \ 24$ kJ mol^{-1}

Examples VII

1. $\Delta S^\circ = 92.2$ J K^{-1} mol^{-1}

2. $\Delta S^\circ = -7.2$ J K^{-1} mol^{-1}
    S° (grey tin) $= 44.3$ J K^{-1} mol^{-1}

3. $\Delta S^\circ = 7.29$ J K^{-1} mol^{-1}

4. $\Delta S^\circ = -174.4$ J K^{-1} mol^{-1}

5. $\Delta S^\circ = 30.9$ J K^{-1} mol^{-1}

6. (a) $\Delta G^\circ \quad = 32.0$ kJ mol^{-1}
    (b) $K_p/p^\circ = 2.14 \times 10^{-2}$

7. (a) $K_p/p^\circ = 1.63 \times 10^{-4}$
    (b) $p_{Zn} \quad = 1.29 \times 10^3$ N m^{-2}

8. $\Delta G^\circ = -75.6$ kJ mol^{-1}

9. $y_A \ = 0.236$

10. $G_A \ = 93.4$ kJ mol^{-1}

11. $K = \dfrac{x^3}{(2 + x)(1-x)^2}$

12. $K = \dfrac{256\ x^4}{(1 + 3x)^3(1-x)}$

13. $x = 0.358$

14. $K_{p/p^\circ} = \dfrac{4\ x^2\ (4-2x)^2\ (p^\circ)^2}{27\ (1-x)^4\ p^2}$

15. $p_{CO} = 11.7\ \text{N m}^{-2}$

16. (a) $\Delta G^\circ = 26\ \text{kJ mol}^{-1}$
    (b) $K_{p/p^\circ} = 2.77 \times 10^{-5}$

17. $\Delta H^\circ = 181\ \text{kJ mol}^{-1}$

18. $\Delta H^\circ = 229\ \text{kJ mol}^{-1}$

19.   (i) c, e, g.
     (ii) a, h.
    (iii) b, d, f.

20.   (i) b.
     (ii) a, c.

Examples VIII

1. (a) $k_3 = \dfrac{1}{2t}\left\{\dfrac{1}{(a-x)^2} - \dfrac{1}{a^2}\right\}$

    (b) Plot $\dfrac{\text{mol}^2\ \text{m}^{-6}}{(c_{Cl_2})^2}$ against t/s.

2. (a) $1.1 \times 10^{-3}\ \text{s}^{-1}$
    (b) 625 s

3. $k_2 = 7.1 \times 10^{-5}\ \text{mol}^{-1}\ \text{m}^3\ \text{s}^{-1}$

4. 0.076

5. 2 400 s

6. $k = 6.65 \times 10^{-3}\ \text{s}^{-1}$

7. $k = 1.94 \times 10^{-4}\ \text{mol}^{-1}\ \text{m}^3\ \text{s}^{-1}$

8. $(CH_3)_3CCl \longrightarrow (CH_3)_3C^+ + Cl^-$ (slow)
    $(CH_3)_3C^+ + H_2O \longrightarrow (CH_3)_3COH + H^+$ (fast)

9. $CH_3COCH_3 + B \longrightarrow CH_3COCH_2^- + HB^+$ (slow)
    $CH_3COCH_2^- + I_2 \longrightarrow CH_3COCH_2I + I^-$ (fast)
    $HB^+ \longrightarrow B + H^+$ (fast)

10. $E = 55.7$ kJ mol^{-1}

11. $E = 97.1$ kJ mol^{-1}

12. $E = 69.7$ kJ mol^{-1}

Examples IX

1. $\pi \;\; = 3.55 \times 10^3$ N m^{-2}

2. $M_r = 123.4$

3. $\alpha \;\; = 0.833$

4. $\Delta p = 64.4$ N m^{-2}

5. $\Delta p = 5.69$ N m^{-2}

6. $M \;\; = 0.161$ kg mol^{-1}

7. $M \;\; = 0.094$ kg mol^{-1}

8. $p \;\; = 5.77 \times 10^4$ N m^{-2}

9. $M \;\; = 0.257$ kg mol^{-1}. The compound is present in the solution as a dimer.

10. $M \;\; = 0.278$ kg mol^{-1}

11. $M \;\; = 0.257$ kg mol^{-1} The molecular formula is S$_8$

12.   0.0667 kg

13. 7.76×10^2 N m^{-2}

14.   Three ions

15. $n = 2$

16.   (a) 6.0 g
     (b) 6.7 g

17.   7.3 g

18. $p = 1.16 \times 10^4$ N m^{-2}

19. $p = 5.75 \times 10^4$ N m^{-2}

20.   53.2%.

Examples X

1.   (a) 1.30×10^{-8} mol kg^{-1}
     (b) $a_{Ag^+} = a_{I^-} = 1.30 \times 10^{-8}$
     (c) $K_s \;\; = 1.69 \times 10^{-16}$

2. $K_S = 5.06 \times 10^{-29}$

3. $s = 1.92 \times 10^{-6}$

4. $s = 2.44 \times 10^{-8}$

5. $s = 4.97 \times 10^{-9}$

6. $s = 1.16 \times 10^{-11}$

7. (a) HSO_3^-, (b) $H_2PO_3^-$, (c) PH_3, (d) $C_6H_5O^-$, (e) C_2H_5OH, (f) $Cr(H_2O)_5OH^{2+}$.

8. (a) $C_3H_7NH_3^+$, (b) HNO_3, (c) $HCrO_4^-$, (d) $(C_2H_5)_2OH^+$, (e) $Al(H_2O)(OH)_2^+$

9. (a) CH_3COO^-, (b) $CH_3COOH_2^+$, (c) SO_4^{2-}, (d) H_2.

10. $\alpha = 0.13$

11. (a) $pK_a = 4.96$, (b) $pK_a = -2.43$.

12. $ClO_4^- < HSO_4^- < CCl_3COO^- < H_2PO_4^- < NH_3 < C_2H_5NH_2$

13. $pH = 7.2$

14. $pH = 7.81$

15. $pH = 2.95$

16. $pH = 9.73$

17. 1.25 g

18. 12.5 mg

19. (a) 6.3, (b) 0.063.

Examples XI

1. (a) $+0.71$ V, (b) $+0.29$ V, (c) $+0.40$ V, (d) $+1.32$ V.

2. I^-, Ag and Mn^{2+} can be oxidised, the last species only to MnO_2.

3. Only Br_2.

4. (a) $5\ Fe^{2+} + MnO_4^- + 8\ H^+ = 5\ Fe^{3+} + Mn^{2+} + 4\ H_2O$
   (b) $2\ I^- + 2\ Fe^{3+} \qquad\quad = I_2 + 2\ Fe^{2+}$
   (c) $HNO_2 + MnO_2 + H^+ \quad = NO_3^- + Mn^{2+} + H_2O$
   (d) $Zn + Ni^{2+} \qquad\qquad = Zn^{2+} + Ni$

5. (a) -357 kJ mol^{-1}
   (b) -44.4 kJ mol^{-1}
   (c) -56.0 kJ mol^{-1}
   (d) -98.4 kJ mol^{-1}

6. (a) $K = 6.4 \times 10^9$
    (b) $K = 1.1 \times 10^3$
    (c) $K = 2.6 \times 10^5$

7. (a) -0.82 V
    (b) $+1.38$ V
    (c) $+0.18$ V

8. (a) -0.029 V
    (b) $+0.18$ V
    (c) $+0.059$ V

9. $a_2 = 1.69$
    pH $= 8.8$

Examples XII

1. 65.8 mg

2. 6.294 mg

3. $\lambda = 1.073 \times 10^{-2}\ \Omega^{-1}\ m^2\ mol^{-1}$

4. $\rho = 0.116\ \Omega\ m$
    $\kappa = 8.63\ \ \Omega^{-1}\ m^{-1}$

5. $l/a = 3.62\ m^{-1}$

6. $\lambda = 1.38 \times 10^{-2}\ \Omega^{-1}\ m^2\ mol^{-1}$

7. $\lambda_0 = 1.145 \times 10^{-2}\ \Omega^{-1}\ m^2\ mol^{-1}$

8. $\lambda_0 = 1.268 \times 10^{-2}\ \Omega^{-1}\ m^2\ mol^{-1}$

9. $\lambda_0 = 40.53 \times 10^{-3}\ \Omega^{-1}\ m^2\ mol^{-1}$

10. $c_{AgCl} = 8.1 \times 10^{-3}\ mol\ m^{-3}$

11. $c_{PbSO_4} = 0.066\ mol\ m^{-3}$

12. $t_{Na^+} = 0.39$

13. $\lambda = 5.87 \times 10^{-3}\ \Omega^{-1}\ m^2\ mol^{-1}$

14. $c_{NaOH} = 22.0\ mol\ m^{-3}$

15. $K = 1.81 \times 10^{-2}\ mol\ m^{-3}$

16. $\alpha = 0.281$

Examples XIII

1. 43% ionic

2. 5.4% ionic

3. $\chi_O = 3.50$

4. $\chi_S = 2.50$

5. $\chi_P = 2.06$

6. 7% ionic

7. 56% ionic

8. In the p-dichloro-compound the two C-Cl bonds are exactly opposite and cancel. The C-Cl bonds are in the same plane as the aromatic ring. In the p-dihydroxy-compound the C-O-H angles are not 180° and the O-H bonds are not in the plane of the ring.

9.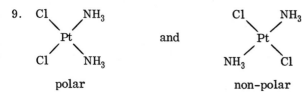

       polar   non-polar

Examples XIV

1. $c(Na_2CO_3) = 213.8$ mol m^{-3}

2. $c(NaOH) = 251.1$ mol m^{-3}

3. $M = 0.3814$ kg mol^{-1}, $x = 10$

4. (a) 50.0 mol m^{-3}, (b) 125.0 mol m^{-3}

5. (a) 21.0 mol m^{-3}, (b) 56.0 mol m^{-3}

6. $w(NH_4Cl) = 102$ mg, $w(NaCl) = 98$ mg

7. (a) +6, (b) -2, (c) +2, (d) +1, (e) +4, (f) +2, (g) +2.5.

8. $c(KMnO_4) = 19.6$ mol m^{-3}

9. $c(Fe^{2+}) = 80.0$ mol m^{-3}, $c(Fe^{3+}) = 48.0$ mol m^{-3}

10. $c(KMnO_4) = 20.0$ mol m^{-3}

11. $c(H_2C_2O_4) = 47.8$ mol m^{-3}, $c(H_2SO_4) = 106.2$ mol m^{-3}

12. $c(H_2O_2) = 20.0$ mol m^{-3},

13. 30.9%

14. $c(KIO_3) = 15.3$ mol m^{-3}

15. $c(NaCl) = 33.5$ mol m^{-3}, $c(HCl) = 42.5$ mol m^{-3}

16. 0.703 g NaCl

17. c (H_2S) $= 19.5$ mol m^{-3}

18. V (KCNS) $= 27.5$ cm^3

19. n_{KClO_3} $= 3.36 \times 10^{-3}$ mol

20. 74.5% Fe, 25.5% Fe_2O_3.

Examples XV

1. C_2H_6O

2. $C_{13}H_{12}N_2S$

3. $C_3H_5O_2$

4. C_3H_8

5. 16.2%

6. Either $CH_3.CH_2.CH_2.CHO$ or $(CH_3)_2CH.CHO$

7. 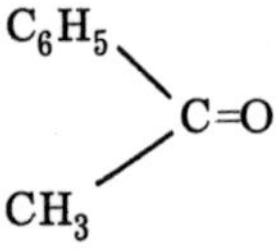 C_6H_5
    \
     $C=O$
    /
    CH_3

8. $C_6H_8O_7$

9. $C_6H_5.NH_2$

10. A, $CH_2Cl.COCl$
     B, $CH_2Cl.COOH$
     C, $CH_2OH.COOH$

11. X, $C_2H_5CONH_2$
     Y, C_2H_5CN
     Z, C_2H_5COOH

12. A, $C_6H_5.O.CO.CH_2.CH_2.CO.OC_6H_5$
     B, C_6H_5OH
     C, CH_2COOH
        |
        CH_2COOH
     D, $CH_2.COOCH_3$
        |
        $CH_2.COOCH_3$
     E, $CH_2.CO$
        | $>O$
        $CH_2.CO$

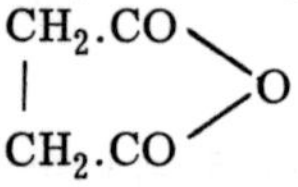

13. X, CH_2OH
        |
        CH_2OH

     Y, CH_2OCOCH_3
        |
        CH_2OCOCH_3

14. (a) 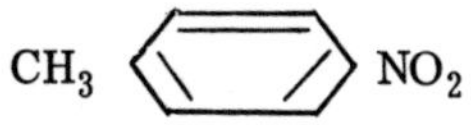 CH_3 ⟨⟩ NO_2

     (b) $CH_3.C_6H_4NO_2 + 6 H^+ + 6 e = CH_3C_6H_4.NH_2 + 2 H_2O$
     (c) $CH_3.C_6H_4.NH_2 + HCl \quad = CH_3.C_6H_4NH_3^+ + Cl^-$
     (d) $CH_3.C_6H_4.NH_3^+ + HNO_2 \quad = CH_3.C_6H_4.N_2^+ + 2 H_2O$
     (e) $CH_3.C_6H_4.N_2^+ + CN^- \quad = CH_3.C_6H_4.CN + N_2$
     (f) $CH_3C_6H_4CN + H^+ + 2 H_2O = CH_3C_6H_4COOH + NH_4^+$

15. X = $\begin{array}{c} NH_2 \\ \\ NH_2 \end{array} \!\! C=O$

     (a) $2 NH_2CONH_2 \qquad = NH_2CONHCONH_2 + NH_3$
     (b) $NH_2CONH_2 + 2 HNO_2 = CO_2 + 3 H_2O + 2 N_2$

16. (a) $COOC_2H_5$
        |
        $COOC_2H_5$

     (b) $COOC_2H_5 \qquad\qquad CONH_2$
        | $\qquad + 2 NH_3 = $ | $\qquad + 2 C_2H_5OH$
        $COOC_2H_5 \qquad\qquad CONH_2$

     (c) $CONH_2 \qquad\qquad COO^{2-}$
        | $\qquad + 2 OH^- = $ | $\qquad + 2 NH_3$
        $CONH_2 \qquad\qquad COO$

     (d) COO^{2-}
        | $\qquad + 2 H^+ \quad = CO + CO_2 + H_2O$
        COO

17. CH_4, 37.7; CO, 43.5; N_2 18.8%.

18. C_3H_8

19. CH_4, 25; C_2H_4, 25; C_2H_2, 50%.

20. C_2H_4, 43; C_3H_8, 57%.

Index

α-Emission, 18
Acceleration, definition, 3
Acid, definition, 133
Acid-base
  indicators, 143
  titrations, 179
Acid dissociation
  constant, 134
  exponent, 135
Activity (radioactive), 21
Ampere, definition, 6
Ampère's law, 4
Analysis
  for carbon and hydrogen, 193
  of mixtures of hydrocarbons, 199
Anion electrode, 147
Anionic acids, 134
Arrhenius' equation, 112
Atomic
  nucleus, 13
  number, 13
  orbitals, 33
  spectrum of hydrogen, 29
  weight, 14
Aufbau principle, 35
Autoprotolysis, 136
  constant for water, 136
Avogadro's constant, 41, 62

β-Emission, 19
Balancing of redox equations, 83
Balmer series, 29, 32
Base, definition, 134
Basic physical quantities, 1
Binding energy of nucleus, 14
Bond dissociation energy, 74
Born exponent, 60, 63
Born-Haber cycle, 64
Born-Landé equation, 59

Boyle's law, 39
Bragg's equation, 55
Buffer
  ratio, 141
  solution, 141
Bunsen's absorption coefficient, 125

Caesium chloride lattice, 56, 58
Capacitance, 171
Capacity factor, 83
Cation electrode, 147
Cationic acids, 134
CGS system, 3
Charles' law, 40
Chemical potential, 89
Coherent systems of units, 3
Colligative properties, 116
  of electrolytes, 120
Concentration
  cells, 155
  definition, 178
Conductivity, 160
  ratio, 166
Conductometric titration, 164
Conjugate acid-base pair, 134
Co-ordination number in crystal, 56
Coulomb, definition, 6
Coulomb's law, 4
Critical
  pressure, 50
  temperature, 50
  volume, 50
Cryoscopic constant, 120

Dalton's law of partial pressures, 44
Daughter nuclide, 19
de Broglie's equation, 28

Decay constant, 20
Degree of ionisation, 121, 165
Depression of freezing point, 120
Derived physical quantities, 1
Dimensionality, 1
Dipole moment, 172
Dissociation of gases, 91
Distribution coefficient, 122

Ebullioscopic constant, 118
Effective nuclear charge, 35
Einstein's equation, 15
Electric
  charge, definition, 3
  potential, definition, 3
Electrolytic
  conduction, 158
  conductivity, 161
Electromagnetic waves, 26
Electromotive force, 148
Electron
  affinity, 65
  as a wave, 28
  cloud, 13, 26
  configurations, 34
Electronegativity, 174
  scales, 174
Electronic promotion energies, 34
Elevation of boiling point, 118
Empirical formulae, 197
Energy of activation, 111
Enthalpy
  of neutralisation, 77
  of reaction, 71
  of solution, 77
Entropy, 82
  of fusion, 85
  of vaporisation, 85
Equilibria
  in gaseous mixtures, 89
  in heterogeneous systems, 95
Equilibrium constant, 81

Exothermic reactions, 70

Faraday's
  constant, 159
  laws of electrolysis, 158
First law of thermodynamics, 72
First-order reaction, 102
Force, definition, 3
Frequency of a vibration, definition, 26

Gas constant, 41
Gibbs function, 88
Graham's law of diffusion, 42

Half-equations, 183
Half-life, 21
  of a reaction, 109
Heading of tables of data, 10
Henry's law, 124
Hess' law of heat summation, 73
Hydrogen ion exponent, 138
Hydrolysis of salts, 140

Ideal gas equation, 9, 41
Indicator constant, 143
Inductive effect, 136
Intensity factor, 83
Iodine titrations, 186
Ionic
  character in bonds, 172
  mobility, 160
  radii, 62
  velocity, 160
Ionisation energy, 33
Isobaric nuclides, 19
Isotonic solutions, 116
Isotopic mass, 14

Joule, definition, 6

Kelvin, definition, 6

Kilogram, definition, 6
Kinetic theory of gases, 44
Kinetics of reaction, 101
Kjeldahl's method of determining nitrogen, 194
Kohlrausch's law of independent mobilities, 162

Labelling of axes in graphs, 11
Lattice energy, 57
Le Chatelier principle, 97
Logarithmic equations, 10
Lyman series, 30, 32

Madelung constant, 60, 62
Mass number, 13
Mean bond energy, 75
Mechanical work, definition, 3
Metallic conduction, 158
Metre, definition, 6
MKS system, 3
Molality and molarity, 207
Molar
  conductivity, 160
    at zero dilution, 161
  mass, definition, 42
Mole
  definition, 6
  fraction, definition, 90, 117
  of reaction, definition, 71
Molecularity of reaction, 109
Molecules in a gas, 44

Nernst's distribution law, 122
Neutron, 13
Newton, definition, 6
Nuclear
  fission, 16
  fusion, 17
Nucleon, 13
Nuclide, 13
Numbers, printing of, 2

Ohm, definition, 6

Osmotic
  pressure, 116
  properties, 116
Ostwald's dilution law, 166
Oxidation, 146
  number, 182
Oxygen-flask technique of analysis, 195

Paschen series, 30, 32
Pauli's exclusion principle, 34
Periodic table, 206
Permanent dipoles, 172
Permeability, 4
Permittivity, 4, 171
pH, 138
Photoelectric effect, 26
Photon, 27
Planck's constant, 27
Polarisation of molecules, 171
Polarity of a bond, 176
Precipitation titrations, 187
Prefixes for fractions and multiples, 7
Protolysis, 134
Proton, 13

Quantisation of energy, 31
Quantity calculus, 8

Radiocarbon dating, 22
Radius ratio rule, 60
Raoult's law, 117
  applied to mixtures, 124
Rate constant of reaction, 102
Rate-determining process, 111
Rate of reaction, 101
Reaction isotherm, 90
Redox
  couple, 148
  processes, 146
  titrations, 182
Reduction, 146

Relative activity
  of an ion in solution, 130
  of a substance, 89
Relative atomic mass, 14
Relative permittivity, definition, 171
Resistivity, definition, 160
Resonance energy, 76
Reversibility, 84, 147
Reversible
  chemical cells, 147
  electrodes, 147
Root mean square velocity, 45
Rydberg constant, 30

Screening constants, 36
Second, definition, 6
Second law of thermodynamics, 82, 86
Second-order reaction kinetics, 101, 105
Semipermeable membrane, 116
Single-bond covalent radii, 175
Slater's rules, 36
Sodium chloride lattice, 56, 57
Solubility
  definition, 130
  product, 131

Standard
  emf, 149
  enthalpy of formation, 72
  hydrogen electrode, 148
  redox potential, 149

Third law of thermodynamics, 85
Threshold frequency, 27
Transport number, 163

Unified atomic mass constant, 14

van der Waals' equation, 48
van't Hoff
  equation, 96
  factor, 121
Velocity, definition, 3
Volt, definition, 6

Wave number, definition, 26
Wavelength, meaning of, 26
Weak electrolytes, 165
Work function, 27
Wurtzite lattice, 57, 59

X-ray analysis of crystals, 55

Zero-order reaction, 103